FLORIDA GEOMETRY EOC TEST PREP 2025 – 2026

All-in-One Study Guide Featuring 400+ Practice Questions and Thorough Answer Explanations for the Florida Geometry EOC

WHITE E. ALLEN

Disclaimer

This study guide is intended to support students in their preparation for the Florida Geometry End-of-Course (EOC) Assessment. While every effort has been made to ensure the accuracy and completeness of the information presented, the authors and publishers make no guarantees regarding the content, scoring, or policies of the official EOC exam.

TABLE OF CONTENTS

INTRODUCTION

Overview of the Florida Geometry EOC Exam

The Florida Geometry End-of-Course (EOC) Assessment is a comprehensive examination designed to measure student achievement of the Florida Standards in Geometry. This standardized assessment serves as both a measure of student learning and a graduation requirement for Florida high school students. Success on this examination demonstrates mastery of essential geometric concepts that form the foundation for more advanced mathematical studies and real-world applications.

The Geometry EOC was established as part of Florida's statewide assessment program to ensure all students graduate with the necessary mathematical knowledge and problem-solving skills for future academic and professional success. The assessment evaluates students' understanding of geometric concepts, their ability to apply these concepts to solve problems, and their capacity to reason mathematically.

This study guide has been meticulously designed to address every standard assessed on the Florida Geometry EOC. Each chapter focuses on specific content areas and provides comprehensive coverage of the concepts, skills, and problem-solving techniques you will need to succeed on the examination. By working through this guide systematically, you will develop both the content knowledge and test-taking strategies necessary to achieve a high score.

Test Format and Structure

The Florida Geometry EOC is administered via computer in most testing situations. The assessment consists of 64-68 questions divided into two sessions, with a brief break between sessions. Students are allocated a total of 160 minutes to complete the entire examination, which is sufficient time for most students to answer all questions and review their work.

The question types on the Geometry EOC include:

1. Multiple-Choice Questions: These questions provide four answer options, only one of which is correct. These questions assess conceptual understanding, procedural fluency, and problem-solving ability.

2. Gridded Response Questions: These questions require students to calculate a numerical answer and enter it in a grid format. No answer choices are provided, so students must determine the answer independently.

3. Multi-Select Questions: These questions ask students to select multiple correct answers from among several options. Students must identify all correct answers to receive full credit.

4. Hot Spot Questions: These questions require students to select one or more "hot spots" on a graphic or figure as their response.
5. Equation Editor/Drag-and-Drop Items: These questions require students to create equations or expressions using a built-in equation editor or by dragging and dropping terms into the correct positions.

The assessment covers content in five major reporting categories:
1. Congruence, Similarity, Right Triangles, and Trigonometry (46% of the test)
2. Circles, Geometric Measurement, and Geometric Properties with Equations (38% of the test)
3. Expressing Geometric Properties with Equations (9% of the test)
4. Modeling with Geometry (7% of the test)

These reporting categories align with the Florida Mathematics Standards for Geometry. The percentage allocations indicate the approximate number of questions you can expect from each category, helping you prioritize your study efforts.

Scoring System and Requirements

The Florida Geometry EOC uses a scaled scoring system ranging from 1 to 5, with Level 3 considered a passing score.

The scale scores correspond to achievement levels as follows:

Achievement Level	Scale Score Range	Performance Description
Level 5	473-575	Mastery - Highly proficient in geometric concepts
Level 4	434-472	Proficient - Above satisfactory understanding
Level 3	399-433	Satisfactory - Passing score that meets graduation requirement
Level 2	368-398	Below Satisfactory - Some understanding but below passing
Level 1	325-367	Inadequate - Limited understanding of concepts

For graduation purposes, students must achieve a Level 3 or higher on the Geometry EOC, or earn a comparable score on an approved alternative assessment. The Geometry EOC score contributes 30% to the student's final course grade in Geometry, making it a significant component of overall academic performance in the course.

It is important to note that students who do not achieve a passing score on their first attempt may retake the assessment during subsequent administration windows. However, thorough preparation using resources like this study guide significantly increases the likelihood of success on the first attempt.

Importance of the Exam

The Geometry EOC serves as a graduation requirement for Florida high school students. Achieving a passing score (Level 3 or higher) fulfills one of the state's assessment requirements for earning a standard high school diploma. Failure to achieve a passing score will necessitate retaking the examination or earning a comparable score on an approved alternative assessment.

Beyond graduation requirements, the Geometry EOC score contributes 30% to a student's final course grade in Geometry. This substantial weight means that EOC performance can significantly impact your GPA and academic standing. A strong performance on the EOC can elevate your overall Geometry grade, potentially strengthening your academic record for college applications.

The knowledge and skills assessed on the Geometry EOC form a foundation for more advanced mathematical studies, including Algebra 2, Precalculus, and Calculus. Mastery of geometric concepts is essential for success in these higher-level courses, which are often required or recommended for competitive college admissions.

Many college STEM programs (Science, Technology, Engineering, and Mathematics) require strong geometric reasoning skills. The concepts assessed on the Geometry EOC—spatial reasoning, proof writing, coordinate geometry, and trigonometry—are fundamental to fields such as engineering, architecture, computer science, and physics.

Success on the Geometry EOC demonstrates your ability to think logically, solve complex problems, and apply mathematical concepts to real-world situations—skills highly valued by both colleges and employers. These transferable skills extend far beyond the geometry classroom.

How to Use This Guide Effectively

This study guide has been specifically designed to optimize your preparation for the Florida Geometry EOC. To derive maximum benefit from this resource, consider the following approach to using the guide effectively.

Begin by familiarizing yourself with the table of contents and overall structure of the guide. Each chapter corresponds to a major content area on the Geometry EOC, with sections organized to build knowledge progressively. Understanding this organization will help you navigate the guide efficiently and create a structured study plan.

The guide employs consistent notation systems throughout to ensure clarity and precision. Mathematical notation in this guide follows standard conventions:

Angles are denoted with the symbol $\angle$, as in $\angle ABC$. Segments are indicated with a bar above the endpoints, as in $\overline{AB}$. Congruent figures are marked with the symbol $\cong$, as in $\triangle ABC \cong \triangle DEF$. Similar figures are indicated with the symbol $\sim$, as in $\triangle ABC \sim \triangle DEF$. Parallel lines are denoted with the symbol $\parallel$, as in $\overline{AB} \parallel \overline{CD}$. Perpendicular lines are indicated with the symbol $\perp$, as in $\overline{AB} \perp \overline{CD}$.

Before beginning intensive study, assess your current knowledge by taking the diagnostic pre-test in Appendix A. This will help identify your strengths and weaknesses, allowing you to prioritize areas needing the most attention. Use the results to create a personalized study plan emphasizing your areas of greatest need.

For each chapter, begin by reading the overview section to establish context for the content. Study the concepts and examples thoroughly, working through each step of the example problems. After studying a section, attempt the practice problems to reinforce your understanding and identify areas requiring further review.

The guide includes numerous practice problems of varying difficulty levels, all designed to mirror the types of questions you will encounter on the actual EOC. Work through these problems systematically, checking your answers against the provided solutions. For incorrect answers, carefully review the solution process to understand your errors.

Study Plan

Effective preparation for the Geometry EOC requires a systematic approach to studying. Begin your preparation early, ideally at least 8-10 weeks before the examination date. This allows sufficient time to cover all content areas thoroughly and provides opportunities for multiple review sessions. Last-minute cramming is particularly ineffective for geometry, which requires deep conceptual understanding rather than mere memorization.

Develop a structured study schedule that allocates specific periods for each content area. A sample study schedule might appear as follows:

Week	Content Focus	Activities
1-2	Foundations & Logic	Study chapters 1-3, complete all practice problems
3-4	Transformations & Congruence	Study chapters 4-5, complete all practice problems
5-6	Polygons & Right Triangles	Study chapters 6-7, complete all practice problems
7-8	Circles & Coordinate Geometry	Study chapters 8-9, complete all practice problems
9	Measurement & Dimension	Study chapter 10, complete all practice problems

| 10 | Comprehensive Review | Take practice tests, review weak areas |

When studying geometric concepts, focus on understanding the underlying principles rather than memorizing procedures. Geometry is highly visual and conceptual, requiring spatial reasoning and logical thinking. Draw diagrams, manipulate figures, and verbalize relationships to deepen your understanding.

Active learning strategies are particularly effective for geometry. These include:

1. Creating your own diagrams and visual representations
2. Teaching concepts to others (or explaining them aloud to yourself)
3. Writing your practice problems and solutions
4. Creating concept maps connecting related geometric ideas
5. Applying concepts to real-world situations or problems

Regular practice with EOC-style questions is essential for success. This guide provides numerous practice opportunities, but consider supplementing with additional practice tests if needed. Simulate test conditions when taking practice tests—use a timer, avoid interruptions, and complete the test in one sitting.

As the examination date approaches, focus on reviewing key concepts rather than learning new material. Create summary sheets of essential formulas, theorems, and procedures for quick reference. Identify and address any remaining weak areas through targeted study.

The night before the examination, avoid intensive studying, which can increase anxiety and fatigue. Instead, review your summary sheets briefly, prepare your testing materials (calculator, pencils, etc.), and ensure you get adequate rest. A well-rested mind performs better on tasks requiring complex thinking and problem-solving.

The preparation for the Geometry EOC is about developing mathematical knowledge and skills that will serve you throughout your academic and professional life. Approach your study with this broader perspective in mind, and you will find the process more meaningful and productive.

FOUNDATIONS OF GEOMETRY

Points, Lines, Planes, and Angles

Geometry begins with undefined terms that serve as the foundation for all geometric concepts. These fundamental elements—points, lines, and planes—cannot be defined using simpler terms, yet they form the basis for defining all other geometric objects and relationships.

A point represents a precise location in space without dimension. We denote points using capital letters. A point has no length, width, or height—it is simply a position. Though we represent points visually as dots, these dots are merely symbols for the conceptual points they represent.

points A, B, and C

A line represents a straight path that extends infinitely in opposite directions. Lines have length but no width or height. We denote lines using lowercase letters or by naming two points on the line. For example, line AB (written as $\overleftrightarrow{AB}$) represents the line passing through points A and B, extending infinitely in both directions.

A line segment is a portion of a line consisting of two endpoints and all points between them. We denote line segments with a bar above the letters naming the endpoints. For example, segment AB (written as $\overline{AB}$) represents the part of the line between points A and B, including A and B.

A ray is a portion of a line consisting of one endpoint and all points extending infinitely in one direction. We denote rays with a half-arrow above the letters naming the endpoint and another point on the ray. For example, ray AB (written as $--\!>$ represents the portion of the line starting at point A and extending infinitely through point B.

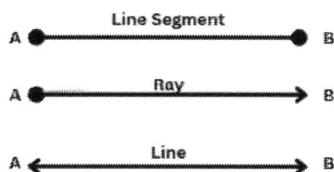

Line, Ray and Line segment

6

A plane represents a flat surface that extends infinitely in all directions. We denote planes using capital script letters (e.g., Plane P) or by naming three non-collinear points that determine the plane. A plane has length and width but no height.

These undefined terms allow us to define more complex geometric objects and relationships. For instance, we can define a circle as the set of all points in a plane that are equidistant from a given point called the center.

Angles form when two rays share a common endpoint called the vertex. We measure angles in degrees, representing the amount of rotation from one ray to another. A complete rotation equals 360 degrees.

We classify angles based on their measure:
- An acute angle measures between 0° and 90° (exclusive)
- A right angle measures exactly 90°
- An obtuse angle measures between 90° and 180° (exclusive)
- A straight angle measures exactly 180°
- A reflex angle measures between 180° and 360° (exclusive)

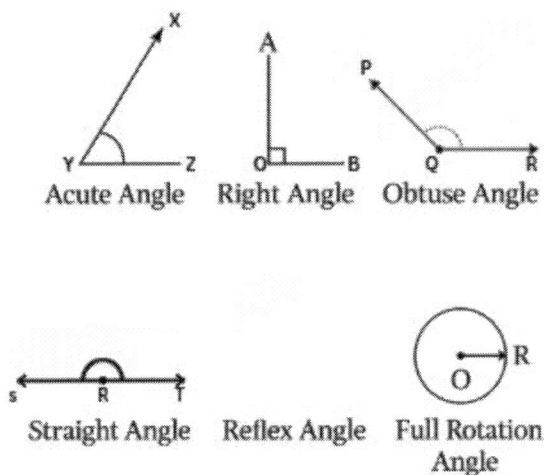

Acute Angle   Right Angle   Obtuse Angle

Straight Angle   Reflex Angle   Full Rotation Angle

Different types of angles

When two lines intersect, they form vertical angles (opposite angles) that are congruent. Adjacent angles formed by intersecting lines are supplementary, meaning they sum to 180°.

Two lines perpendicular to each other form right angles at their intersection. We denote perpendicular lines using the symbol ⊥. For example, AB⊥CD indicates that lines AB and CD are perpendicular.

Parallel lines never intersect and maintain the same distance from each other throughout their infinite extension. We denote parallel lines using the symbol ∥. For example, AB∥CD indicates that lines AB and CD are parallel.

When a transversal intersects parallel lines, it creates eight angles with special relationships. Corresponding angles are congruent, alternate interior angles are congruent, and alternate exterior angles are congruent. Consecutive interior angles and consecutive exterior angles are supplementary.

Coordinate Systems

The Cartesian coordinate system consists of two perpendicular number lines that intersect at their origins. The horizontal number line is called the x-axis, and the vertical number line is called the y-axis. Their point of intersection is the origin, denoted as (0, 0).

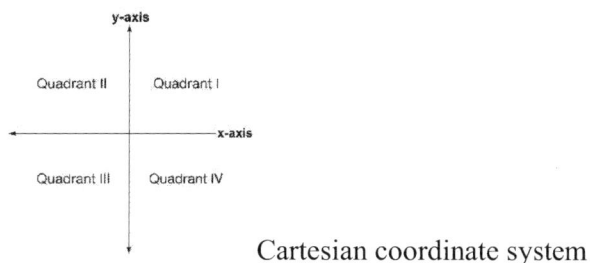

Cartesian coordinate system

The coordinate plane is divided into four quadrants, numbered counterclockwise from the upper right:
- Quadrant I: Both x and y coordinates are positive
- Quadrant II: x-coordinate is negative, y-coordinate is positive
- Quadrant III: Both x and y coordinates are negative
- Quadrant IV: x-coordinate is positive, y-coordinate is negative

Each point in the coordinate plane is identified by an ordered pair (x, y), where x represents the horizontal distance from the origin, and y represents the vertical distance. Points to the right of the y-axis have positive x-coordinates, while points to the left have negative x-coordinates. Similarly, points above the x-axis have positive y-coordinates, while points below have negative y-coordinates.

To plot a point, we start at the origin and move horizontally according to the x-coordinate, then vertically according to the y-coordinate. For example, to plot the point (3, 4), we move 3 units right from the origin, then 4 units up.

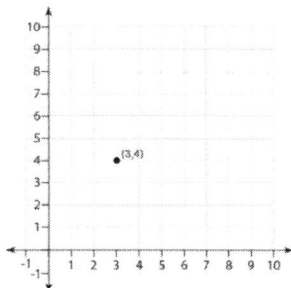

8

The coordinate system allows us to describe geometric figures algebraically. For instance, a line can be represented by a linear equation in the form $ax + by + c = 0$, and a circle can be represented by an equation in the form $(x - h)^2 + (y - k)^2 = r^2$, where (h, k) is the center and r is the radius.

The coordinate system also enables us to verify geometric properties algebraically. For example, we can determine whether a quadrilateral is a parallelogram by checking if its opposite sides have equal slopes, or whether a triangle is right-angled by applying the Pythagorean theorem to the coordinates of its vertices.

In three-dimensional geometry, we extend the coordinate system to include a third axis, the z-axis, which is perpendicular to both the x and y axes. Each point in three-dimensional space is identified by an ordered triple (x, y, z).

Distance and Midpoint

The distance formula derives from the Pythagorean theorem. For two points (x_1, y_1) and (x_2, y_2) in the coordinate plane, the distance between them is given by:

$$d = \sqrt{(x_2 - x_1)^2 + (y_2 - y_1)^2}$$

To derive this formula, we can create a right triangle with the line segment connecting the two points as the hypotenuse. The legs of the right triangle have lengths $|x_2 - x_1|$ and $|y_2 - y_1|$. Applying the Pythagorean theorem, the square of the hypotenuse equals the sum of the squares of the legs:

$$d^2 = |x_2 - x_1|^2 + |y_2 - y_1|^2 = (x_2 - x_1)^2 + (y_2 - y_1)^2$$

Taking the square root of both sides gives us the distance formula.

Example: Find the distance between points A(3, -2) and B(-4, 5).

Solution: $d = \sqrt{(-4 - 3)^2 + \left(5 - (-2)\right)^2} = \sqrt{(-7)^2 + (7)^2} = \sqrt{49 + 49} = \sqrt{98} = 7\sqrt{2} \approx 9.90$

The midpoint formula gives us the coordinates of the point exactly halfway between two given points. For two points (x_1, y_1) and (x_2, y_2), the midpoint M is given by:

$$M = \left(\frac{x_1 + x_2}{2}, \frac{y_1 + y_2}{2}\right)$$

This formula represents the average of the x-coordinates and the average of the y-coordinates of the two endpoints.

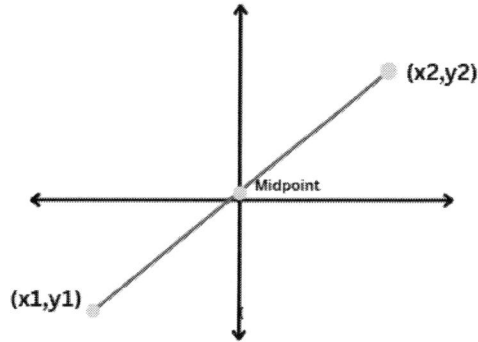

Midpoint M of segment AB with coordinates

Example: Find the midpoint of the line segment with endpoints $C(-6, 8)$ and $D(10, -4)$.

Solution:

$$M = \left(\frac{-6 + 10}{2}, \frac{8 + (-4)}{2}\right) = \left(\frac{4}{2}, \frac{4}{2}\right) = (2,2)$$

The distance and midpoint formulas have numerous applications in coordinate geometry. For instance, we can use the distance formula to verify that a triangle is isosceles by showing that two sides have equal lengths, or to confirm that a quadrilateral is a rhombus by demonstrating that all four sides have equal lengths.

Similarly, we can use the midpoint formula to find the coordinates of the center of a circle given the coordinates of the endpoints of a diameter, or to locate the centroid of a triangle by finding the midpoint of each side.

In three-dimensional space, these formulas extend naturally. The distance between points (x_1, y_1, z_1) and (x_2, y_2, z_2) is given by:

$$d = \sqrt{(x_2 - x_1)^2 + (y_2 - y_1)^2 + (z_2 - z_1)^2}$$

And the midpoint is:

$$M = \left(\frac{x_1 + x_2}{2}, \frac{y_1 + y_2}{2}, \frac{z_1 + z_2}{2}\right)$$

Geometric Constructions

Geometric constructions allow us to create precise geometric figures using only a straightedge (unmarked ruler) and compass. These classical construction techniques, dating back to ancient Greek mathematics, provide a foundation for understanding geometric relationships without relying on measurement.

The basic operations in geometric constructions include:
- Drawing a straight line through two points
- Drawing a circle with a given center and radius
- Finding the intersection points of lines and circles

From these basic operations, we can perform more complex constructions. Let's examine some fundamental constructions relevant to the Florida Geometry EOC.

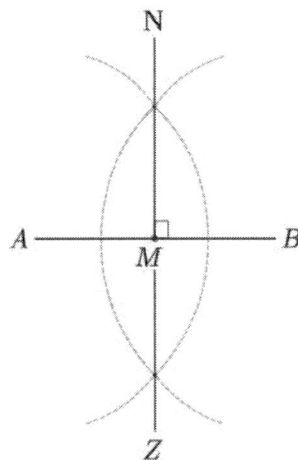

Perpendicular bisector

Construction 1: Perpendicular Bisector of a Line Segment Given a line segment AB, we can construct its perpendicular bisector as follows:

1. Place the compass point at A and set the radius to more than half the length of AB.
2. Draw an arc above and below the line segment.
3. Without changing the compass setting, place the compass point at B and draw arcs that intersect the previous arcs.
4. Draw a line through the two intersection points. This line is the perpendicular bisector of AB.

This construction creates a line that is perpendicular to AB and passes through its midpoint. Every point on this perpendicular bisector is equidistant from points A and B.

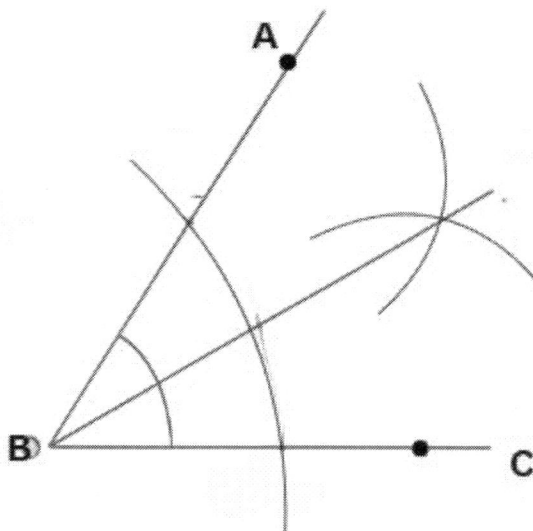

Construction of angle bisector

Construction 2: Angle Bisector Given an angle $\angle ABC$, we can construct its bisector as follows:

1. Place the compass point at B and draw an arc that intersects both sides of the angle.
2. Place the compass point at the intersection of the arc with one side of the angle, and draw an arc in the interior of the angle.
3. Without changing the compass setting, place the compass point at the intersection of the first arc with the other side of the angle, and draw an arc that intersects the previous arc.
4. Draw a ray from B through the intersection point of the two arcs in the interior of the angle. This ray is the angle bisector.

This construction creates a ray that divides the original angle into two congruent angles. Every point on this angle bisector is equidistant from the two sides of the original angle.

Construction 3: Line Parallel to a Given Line Through a Given Point Given a line ℓ and a point P not on ℓ, we can construct a line through P parallel to ℓ as follows:

1. Draw a transversal line through P intersecting ℓ at point Q.
2. Place the compass point at Q and draw an arc that intersects ℓ.
3. Place the compass point at P and draw an arc on the same side of the transversal.
4. Set the compass to the distance between Q and the intersection of the first arc with ℓ.
5. Place the compass point where the second arc intersects the transversal and draw a third arc.
6. Draw a line through P and the intersection of the second and third arcs. This line is parallel to ℓ.

This construction creates a line through P that never intersects ℓ, meeting the definition of parallel lines.

Construction 4: Line Perpendicular to a Given Line Through a Given Point Given a line ℓ and a point P (either on or not on ℓ), we can construct a line through P perpendicular to ℓ as follows:

If P is not on ℓ:

1. Place the compass point at P and draw an arc that intersects ℓ at two points, A and B.

2. Place the compass point at A and draw an arc in the vicinity of P.

3. Without changing the compass setting, place the compass point at B and draw an arc that intersects the previous arc.

4. Draw a line through P and the intersection of the two arcs. This line is perpendicular to ℓ.

If P is on ℓ:

1. Place the compass point at P and draw an arc that intersects ℓ at two points, A and B.

2. Place the compass point at A and draw an arc above ℓ.

3. Without changing the compass setting, place the compass point at B and draw an arc that intersects the previous arc.

4. Draw a line through P and the intersection of the two arcs. This line is perpendicular to ℓ.

Practice Problems

1. Find the distance between points $P(-3, 5)$ and $Q(2, -4)$.

2. Determine the midpoint of the line segment with endpoints $A(7, -2)$ and $B(-5, 8)$.

3. Points M and N are the midpoints of sides AB and AC of triangle ABC, respectively. If $A(3, 7), B(-1, 5)$, and $C(5, 1)$, find the coordinates of M and N.

4. The distance from point $P(x, 5)$ to point $Q(3, -2)$ is 10 units. Find the possible values of x.

5. Triangle ABC has vertices $A(2, 3)$, $B(6, 3)$, and $C(4, 7)$. Verify that this triangle is isosceles by calculating the lengths of its sides.

Solutions

1. Using the distance formula: $d = \sqrt{\left(2 - (-3)\right)^2 + (-4 - 5)^2} = \sqrt{5^2 + (-9)^2} = \sqrt{25 + 81} = \sqrt{106} \approx$ 10.30

2. Using the midpoint formula: $M = \left(\frac{7+(-5)}{2}, \frac{-2+8}{2}\right) = \left(\frac{2}{2}, \frac{6}{2}\right) = (1, 3)$

3. For midpoint M of AB: $M = \left(\frac{3+(-1)}{2}, \frac{7+5}{2}\right) = \left(\frac{2}{2}, \frac{12}{2}\right) = (1, 6)$

   For midpoint N of AC: $N = \left(\frac{3+5}{2}, \frac{7+1}{2}\right) = \left(\frac{8}{2}, \frac{8}{2}\right) = (4, 4)$

4. Using the distance formula: $10 = \sqrt{(x - 3)^2 + \left(5 - (-2)\right)^2} = \sqrt{(x - 3)^2 + 7^2} 100 = (x - 3)^2 + (x - 3)^2 = x - 3 = \pm\sqrt{51} x = 3 \pm \sqrt{51} x \approx 3 + 7.14 \approx 10.14$ or $x \approx 3 - 7.14 \approx -4.1$

5. Finding the lengths of all three sides:

   Length of AB: $AB = \sqrt{(6 - 2)^2 + (3 - 3)^2} = \sqrt{16 + 0} = 4$

13

Length of BC: $BC = \sqrt{(4-6)^2 + (7-3)^2} = \sqrt{4+16} = \sqrt{20} = 2\sqrt{5}$

Length of AC: $AC = \sqrt{(4-2)^2 + (7-3)^2} = \sqrt{4+16} = \sqrt{20} = 2\sqrt{5}$

Since $BC = AC = 2\sqrt{5}$, triangle ABC is isosceles.

LOGICAL REASONING AND PROOF

Conditional Statements

Logical reasoning forms the foundation of geometric proof. At the heart of this reasoning are conditional statements, which express relationships between hypotheses and conclusions. A conditional statement has the form "if p, then q," where p is the hypothesis (or antecedent) and q is the conclusion (or consequent). In mathematical notation, this is written as $p \rightarrow q$.

The truth value of a conditional statement depends solely on the truth values of its component parts. A conditional statement is false only when the hypothesis is true and the conclusion is false. In all other cases, the conditional statement is true. This relationship can be summarized in a truth table:

p	q	$p \rightarrow q$
T	T	T
T	F	F
F	T	T
F	F	T

For example, consider the conditional statement "If a triangle is equilateral, then it is equiangular." This statement is true because all equilateral triangles (triangles with three equal sides) are indeed equiangular (have three equal angles).

Three related statements can be formed from any conditional:

1. The converse: "If q, then p" (written as $q \rightarrow p$)
2. The inverse: "If not p, then not q" (written as $\neg p \rightarrow \neg q$)
3. The contrapositive: "If not q, then not p" (written as $\neg q \rightarrow \neg p$)

For the example "If a triangle is equilateral, then it is equiangular":

1. Converse: "If a triangle is equiangular, then it is equilateral."
2. Inverse: "If a triangle is not equilateral, then it is not equiangular."
3. Contrapositive: "If a triangle is not equiangular, then it is not equilateral."

A key property of conditional statements is that a conditional and its contrapositive are logically equivalent—they always have the same truth value. Similarly, the converse and inverse of a statement are logically equivalent to each other. However, the converse of a true statement is not necessarily true.

In the example above, both the original statement and its converse happen to be true. When both a conditional and its converse are true, the relationship is called a biconditional, written as "p if and only if q" or symbolically as $p \leftrightarrow q$.

Conditional statements often appear in the form of definitions and theorems in geometry. For instance, the definition "A quadrilateral is a rhombus if and only if it has four congruent sides" is a biconditional statement. The "if and only if" indicates that having four congruent sides is both necessary and sufficient for a quadrilateral to be classified as a rhombus.

Inductive and Deductive Reasoning

Mathematics employs two primary forms of reasoning: inductive and deductive. Each plays a distinct role in the development and application of geometric knowledge.

Inductive reasoning involves drawing general conclusions from specific observations or examples. It proceeds from the particular to the general. In geometry, inductive reasoning might involve examining several specific cases and identifying a pattern or general rule.

For example, after calculating the sum of interior angles for several polygons—180° for a triangle, 360° for a quadrilateral, 540° for a pentagon—one might induce the general formula for the sum of interior angles in an n-sided polygon: $(n - 2) \times 180°$.

While inductive reasoning is valuable for discovering patterns and formulating conjectures, it does not constitute proof. The conclusion reached through inductive reasoning remains a conjecture until proven deductively. No matter how many specific cases support a conjecture, a single counterexample can disprove it.

Deductive reasoning, in contrast, involves drawing specific conclusions from general principles or premises. It proceeds from the general to the particular. In geometry, deductive reasoning forms the basis of formal proof.

A deductive argument typically follows the structure:
1. General principles (axioms, postulates, definitions, previously proven theorems)
2. Specific given information (hypotheses)
3. Logical steps of reasoning
4. Conclusion

For example, using the general principles of triangle congruence (such as the SSS postulate) and specific given information about two triangles, one can deductively prove that the triangles are congruent.

Deductive reasoning guarantees the truth of the conclusion if:
1. The general principles used are true, and
2. The logical reasoning steps are valid.

This certainty distinguishes deductive reasoning from inductive reasoning, which yields only probable conclusions.

The axiomatic system of geometry, established by Euclid, exemplifies deductive reasoning. Beginning with a small set of undefined terms (point, line, plane) and axioms (self-evident truths), the entire structure of Euclidean geometry is built through deductive reasoning. Each theorem follows logically from previously established results, creating a coherent and logically consistent system.

Types of Geometric Proofs

Geometry employs several formats for presenting proofs, each with distinct advantages for different types of problems. The three primary proof formats are two-column, paragraph, and flow proofs.

The two-column proof is the most structured format, organizing the logical argument into two parallel columns. The left column contains statements, while the right column provides justification for each statement (reasons). This format clearly displays the logical progression from given information to conclusion, making it especially effective for straightforward proofs where each step directly builds on previous ones.

For example, a two-column proof might establish that if two angles are vertical angles, then they are congruent:

Statement	Reason
1. Angles 1 and 3 are vertical angles	1. Given
2. Angle 1 and Angle 2 form a straight angle	2. Definition of vertical angles
3. $m\angle 1 + m\angle 2 = 180°$	3. Definition of straight angle
4. Angle 2 and Angle 3 form a straight angle	4. Definition of vertical angles
5. $m\angle 2 + m\angle 3 = 180°$	5. Definition of straight angle
6. $m\angle 1 + m\angle 2 = m\angle 2 + m\angle 3$	6. Transitive property of equality (steps 3 and 5)
7. $m\angle 1 = m\angle 3$	7. Subtraction property of equality

| 8. *Angle* 1 ≅ *Angle* 3 | 8. Definition of congruent angles |

The paragraph proof presents the same logical argument in continuous prose. This format often appears more natural and allows for more complex explanations of the reasoning process. Paragraph proofs are particularly useful for more intricate arguments where the connections between steps may require elaboration.

For the same example of vertical angles, a paragraph proof might read:

"Given vertical angles 1 and 3, we know that angles 1 and 2 form a straight angle. By the definition of a straight angle, the sum of their measures is 180°, so $m\angle 1 + m\angle 2 = 180°$. Similarly, angles 2 and 3 form a straight angle, so $m\angle 2 + m\angle 3 = 180°$. Since both sums equal 180°, we can write $m\angle 1 + m\angle 2 = m\angle 2 + m\angle 3$. Subtracting $m\angle 2$ from both sides yields $m\angle 1 = m\angle 3$. Therefore, by the definition of congruent angles, angle 1 is congruent to angle 3."

The flow proof (also called a tree proof) represents the logical structure visually, with statements connected by arrows indicating the flow of logic. Each statement appears in a box, with arrows showing which previous statements contribute to its justification. This format clearly displays the dependencies between statements and can be particularly helpful for visualizing complex logical structures.

Indirect proof, or proof by contradiction, offers an alternative approach when direct proof is challenging. This method assumes the negation of the desired conclusion and demonstrates that this assumption leads to a contradiction of known facts or the given information. Since the assumption leads to an impossibility, it must be false, and therefore the original statement must be true.

For example, to prove that there is no greatest even integer, one would assume there is a greatest even integer, call it n. Then $n + 2$ would be an even integer greater than n, contradicting the assumption that n is the greatest even integer. This contradiction proves that there is no greatest even integer.

The choice of proof format depends on the nature of the problem, the complexity of the logical argument, and the intended audience. Mastery of multiple proof formats provides flexibility in addressing different types of geometric problems.

Proofs Involving Angles and Lines

Proofs involving angles and lines establish key relationships that extend to more complex geometric figures. These proofs typically rely on angle relationships formed by intersecting lines, parallel lines cut by transversals, and the properties of polygons.

One fundamental relationship involves vertical angles, which are formed when two lines intersect. Vertical angles are opposite each other at the intersection and are always congruent. A formal proof of this property was demonstrated in the previous section.

Another important set of relationships emerges when parallel lines are cut by a transversal. Eight angles are formed, with several significant relationships:

- Corresponding angles are congruent
- Alternate interior angles are congruent
- Alternate exterior angles are congruent
- Consecutive interior angles are supplementary (sum to 180°)
- Consecutive exterior angles are supplementary (sum to 180°)

These relationships can be proven using the properties of vertical angles and the parallel postulate.

For example, to prove that alternate interior angles are congruent when parallel lines are cut by a transversal:

Statement	Reason
1. Lines ℓ and m are parallel	1. Given
2. Transversal t intersects lines ℓ and m	2. Given
3. Angle 3 and angle 5 are alternate interior angles	3. Definition of alternate interior angles
4. Angle 3 and angle 7 are corresponding angles	4. Definition of corresponding angles
5. Angle 3 ≅ Angle 7	5. Corresponding angles are congruent when lines are parallel
6. Angle 5 and angle 7 are vertical angles	6. Definition of vertical angles
7. Angle 5 ≅ Angle 7	7. Vertical angles are congruent
8. Angle 3 ≅ Angle 5	8. Transitive property of congruence (steps 5 and 7)

Conversely, if two lines are cut by a transversal and corresponding angles are congruent, then the lines are parallel. This converse relationship provides a means to prove that lines are parallel based on angle relationships.

Angle relationships also play a crucial role in proofs involving polygons. For instance, the sum of interior angles in a triangle equals 180°, which can be proven using the parallel postulate and angle relationships.

Statement	Reason
1. Triangle ABC with vertices A, B, and C	1. Given
2. Draw a line through point A parallel to side BC	2. Through a point not on a line, exactly one line can be drawn parallel to the given line
3. Angle 1 ≅ Angle 4	3. Corresponding angles are congruent when lines are parallel
4. Angle 3 ≅ Angle 6	4. Corresponding angles are congruent when lines are parallel
5. $m\angle 4 + m\angle 5 + m\angle 6 = 180°$	5. Angles on a straight line sum to 180°
6. $m\angle 1 + m\angle 2 + m\angle 3 = 180°$	6. Substitution property (steps 3, 4, and 5)
7. The sum of interior angles in triangle ABC is 180°	7. Angle 1, angle 2, and angle 3 are the interior angles of triangle ABC

This result extends to any polygon. For an n-sided polygon, the sum of interior angles equals $(n-2) \times 180°$, which can be proven by dividing the polygon into triangles.

Parallel Lines and Transversals

Parallel lines never intersect and maintain a constant distance from each other. When a transversal intersects parallel lines, it creates special angle relationships that are fundamental to many geometric proofs.

Consider two parallel lines, ℓ and m, intersected by a transversal t. This configuration creates eight angles, typically numbered as shown in the diagram.

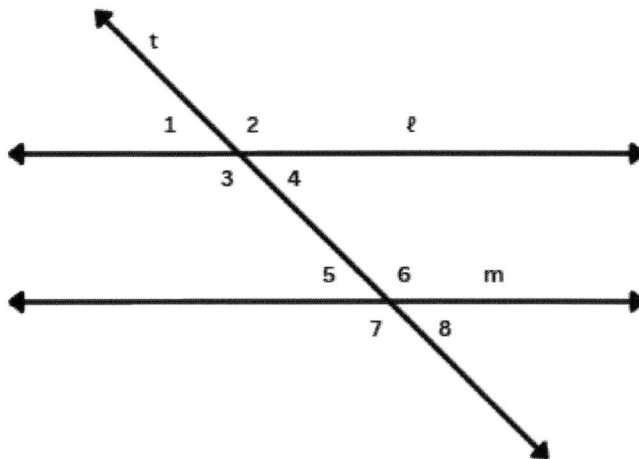

The angle relationships in this configuration include:

Corresponding angles are congruent: $\angle 1 \cong \angle 5$ $\angle 2 \cong \angle 6$ $\angle 3 \cong \angle 7$ $\angle 4 \cong \angle 8$

Alternate interior angles are congruent: $\angle 3 \cong \angle 5$ $\angle 4 \cong \angle 6$

Alternate exterior angles are congruent: $\angle 1 \cong \angle 7$ $\angle 2 \cong \angle 8$

Consecutive interior angles are supplementary: $m\angle 3 + m\angle 6 = 180°$ $m\angle 4 + m\angle 5 = 180°$

Consecutive exterior angles are supplementary: $m\angle 1 + m\angle 8 = 180°$ $m\angle 2 + m\angle 7 = 180°$

These relationships provide a means to find angle measures and also serve as criteria for establishing that lines are parallel. The converses of these relationships are also true:

1. If corresponding angles are congruent, then the lines are parallel.
2. If alternate interior angles are congruent, then the lines are parallel.
3. If alternate exterior angles are congruent, then the lines are parallel.
4. If consecutive interior angles are supplementary, then the lines are parallel.
5. If consecutive exterior angles are supplementary, then the lines are parallel.

These converses provide multiple approaches for proving that lines are parallel based on angle relationships.

For example, to prove that a pair of lines are parallel given that alternate interior angles are congruent:

Statement	Reason
1. Transversal t intersects lines ℓ and m	1. Given
2. Angle 3 ≅ Angle 5	2. Given (alternate interior angles)
3. Lines ℓ and m are parallel	3. If alternate interior angles are congruent, then the lines are parallel

The angle relationships associated with parallel lines cut by a transversal have numerous applications in geometric proofs. They are particularly useful for proving properties of triangles and quadrilaterals.

For instance, the fact that the sum of interior angles in a triangle equals 180° can be proven using these relationships. Similarly, the properties of special quadrilaterals, such as parallelograms, can be established using parallel lines and transversals.

When working with parallel lines and transversals, it's important to identify angle pairs correctly. Corresponding angles occur in the same relative position, alternate angles lie on opposite sides of the transversal, and consecutive angles are adjacent on the same side of the transversal.

Practice Problems

1. Determine whether the following statement is true or false. If false, provide a counterexample: "If two angles are supplementary, then they are adjacent."
2. Write the converse, inverse, and contrapositive of the statement: "If a quadrilateral has four right angles, then it is a rectangle."
3. Prove that the sum of the measures of the interior angles of a quadrilateral is 360°.
4. Given: Line ℓ is parallel to line m, and line n is parallel to line p. Prove that if lines ℓ and n are perpendicular, then lines m and p are perpendicular.

Solutions

1. The statement "If two angles are supplementary, then they are adjacent" is false.

   Counterexample: Consider two angles in different planes or two non-adjacent angles in the same plane whose measures sum to 180°. For instance, in a parallelogram, opposite angles are not adjacent, but pairs of consecutive angles are supplementary.

2. Original statement: "If a quadrilateral has four right angles, then it is a rectangle."

   Converse: "If a quadrilateral is a rectangle, then it has four right angles." This statement is true by the definition of a rectangle.

   Inverse: "If a quadrilateral does not have four right angles, then it is not a rectangle." This statement is true because by definition, rectangles must have four right angles.

   Contrapositive: "If a quadrilateral is not a rectangle, then it does not have four right angles." This statement is true and logically equivalent to the original statement.

3. Proof that the sum of the measures of the interior angles of a quadrilateral is 360°:

Statement	Reason
1. Consider quadrilateral ABCD	1. Given
2. Draw diagonal AC, dividing the quadrilateral into triangles ABC and ACD	2. Through any two non-adjacent vertices of a quadrilateral, a diagonal can be drawn
3. The sum of interior angles in triangle ABC is 180°	3. The sum of interior angles in any triangle is 180°
4. The sum of interior angles in triangle ACD is 180°	4. The sum of interior angles in any triangle is 180°
5. The sum of interior angles in quadrilateral ABCD is 180° + 180° = 360°	5. The sum of angles in both triangles equals the sum of interior angles in the quadrilateral

4. Proof that if lines ℓ and n are perpendicular, and line ℓ is parallel to line m, and line n is parallel to line p, then lines m and p are perpendicular:

Statement	Reason
1. Line ℓ ⊥ line n	1. Given
2. Line ℓ ∥ line m	2. Given
3. Line n ∥ line p	3. Given
4. $m\angle 1 = 90°$ (where angle 1 is formed by lines ℓ and n)	4. Definition of perpendicular lines
5. Line n intersects lines ℓ and m	5. Given
6. Angle 1 and angle 2 (formed by lines n and m) are corresponding angles	6. Definition of corresponding angles
7. $m\angle 2 = m\angle 1 = 90°$	7. Corresponding angles are congruent when lines are parallel
8. Line m ⊥ line n	8. Definition of perpendicular lines
9. Line m intersects lines n and p	9. Given
10. Angle 2 and angle 3 (formed by lines m and p) are corresponding angles	10. Definition of corresponding angles
11. $m\angle 3 = m\angle 2 = 90°$	11. Corresponding angles are congruent when lines are parallel
12. Line m ⊥ line p	12. Definition of perpendicular lines

TRANSFORMATIONS

Translations

A translation is a transformation that moves every point of a figure the same distance in the same direction. This transformation is sometimes described as a "slide" because the figure maintains its size, shape, and orientation while changing position. Translations preserve both distance and angle measure, making them rigid transformations or isometries.

In coordinate geometry, a translation can be represented as a function that maps each point (x, y) to a new point $(x + h, y + k)$, where h represents the horizontal displacement and k represents the vertical displacement. This mapping can be written as:

$$(x, y) \rightarrow (x + h, y + k)$$

The values of h and k determine the direction and magnitude of the translation. If $h > 0$, the figure moves to the right; if $h < 0$, it moves to the left. Similarly, if $k > 0$, the figure moves up; if $k < 0$, it moves down.

For example, under the translation $(x, y) \rightarrow (x + 3, y - 2)$, the point $(1, 4)$ maps to $(1 + 3, 4 - 2) = (4, 2)$. Every point in the figure undergoes the same displacement, moving 3 units right and 2 units down.

The notation $T(h, k)$ is often used to represent a translation where h is the horizontal displacement and k is the vertical displacement. To apply $T(h, k)$ to a point (x, y), calculate:

$$T_{(h,k)}(x, y) = (x + h, y + k)$$

For polygons defined by vertices, apply the translation to each vertex and connect the resulting points to form the image. For example, to translate triangle ABC with vertices $A(1, 2), B(3, 2)$, and $C(2, 5)$ by $T(4, -1)$:

$$A(1, 2) \rightarrow A'(1 + 4, 2 - 1) = A'(5, 1)$$

$$B(3, 2) \rightarrow B'(3 + 4, 2 - 1) = B'(7, 1)$$

$$C(2, 5) \rightarrow C'(2 + 4, 5 - 1) = C'(6, 4)$$

Connecting these translated vertices creates triangle A'B'C', the image of triangle ABC under the translation $T(4, -1)$.

Translations have several key properties:
1. The pre-image and image are congruent figures.
2. Corresponding angles remain equal.

24

3. Corresponding segments remain parallel and equal in length.
4. The orientation of the figure is preserved.
5. The translation of a line is a parallel line.

In geometric constructions, translations can be performed using a compass and straightedge by transferring distances and creating parallel lines. In real-world applications, translations model linear movement, such as the displacement of objects or the sliding of structures.

Reflections

A reflection is a transformation that creates a mirror image of a figure across a line called the line of reflection. Each point of the original figure (pre-image) maps to a point such that the line of reflection is the perpendicular bisector of the segment connecting the original point and its image. Reflections preserve distance and angle measure but reverse orientation, making them rigid transformations or isometries.

The most commonly used lines of reflection in coordinate geometry are the $x-axis$, the $y-axis$, the line $y = x$, and the line $y = -x$. The mapping rules for these standard reflections are:

1. Reflection over the x-axis: $(x, y) \rightarrow (x, -y)$
2. Reflection over the y-axis: $(x, y) \rightarrow (-x, y)$
3. Reflection over the line $y = x$: $(x, y) \rightarrow (y, x)$
4. Reflection over the line $y = -x$: $(x, y) \rightarrow (-y, -x)$

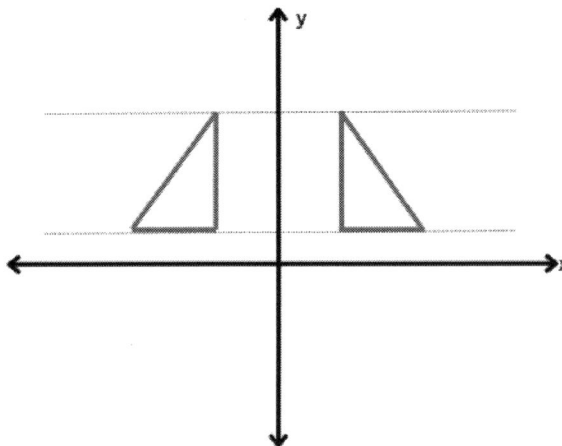

Reflection of a triangle over the y -axis

For a reflection over an arbitrary line with equation $ax + by + c = 0$, the mapping becomes more complex and involves projecting points perpendicular to the line of reflection.

For example, under the reflection over the x-axis, the point $(3, 4)$ maps to $(3, -4)$. Notice that the x-coordinate remains unchanged while the y-coordinate changes sign. Similarly, under the reflection over the y-axis, the point $(3, 4)$ maps to $(-3, 4)$.

The notation r_{x-axis}, r_{y-axis}, $r_{y=x}$, or $r_{y=-x}$ is often used to represent reflections over the respective lines. To apply these reflections to a point (x, y), calculate:

$$r_{x-axis}(x, y) = (x, -y) \quad r_{y-axis}(x, y) = (-x, y) \quad r_{y=x}(x, y) = (y, x) \quad r_{y=-x}(x, y) = (-y, -x)$$

For polygons defined by vertices, apply the reflection to each vertex and connect the resulting points to form the image. For example, to reflect triangle ABC with vertices $A(1, 2)$, $B(3, 2)$, and $C(2, 5)$ over the x-axis:

$$A(1, 2) \rightarrow A'(1, -2)$$

$$B(3, 2) \rightarrow B'(3, -2)$$

$$C(2, 5) \rightarrow C'(2, -5)$$

Connecting these reflected vertices creates triangle A'B'C', the image of triangle ABC under the reflection over the x-axis.

Reflections have several key properties:

1. The pre-image and image are congruent figures.
2. Corresponding angles remain equal.
3. Corresponding segments remain equal in length.
4. The orientation of the figure is reversed (like a mirror image).
5. A point on the line of reflection maps to itself.
6. Two successive reflections over parallel lines result in a translation.
7. Two successive reflections over intersecting lines result in a rotation around the point of intersection.

In geometric constructions, reflections can be performed using perpendicular bisectors. In real-world applications, reflections model mirror images, symmetry in nature and design, and optical phenomena.

Rotations

A rotation is a transformation that turns a figure around a fixed point called the center of rotation. Each point of the original figure (pre-image) moves along a circular arc centered at the center of rotation through the same angle. Rotations preserve distance and angle measure, making them rigid transformations or isometries.

In coordinate geometry, rotations are typically performed around the origin (0, 0) through standard angles. The mapping rules for these standard rotations around the origin are:

1. Rotation by 90° counterclockwise: $(x, y) \rightarrow (-y, x)$
2. Rotation by 180° counterclockwise: $(x, y) \rightarrow (-x, -y)$
3. Rotation by 270° counterclockwise: $(x, y) \rightarrow (y, -x)$

For a rotation by an arbitrary angle θ around the origin, the mapping rule becomes:

$$(x, y) \rightarrow (x\cos\theta - y\sin\theta, x\sin\theta + y\cos\theta)$$

This formula uses the trigonometric functions to calculate the new coordinates based on the angle of rotation.

For example, under the rotation by 90° counterclockwise around the origin, the point (3, 4) maps to (-4, 3). Under the rotation by 180° around the origin, the point (3, 4) maps to (-3, -4).

The notation $R(O, \theta)$ is often used to represent a rotation around point O by angle θ. For rotations around the origin, this can be simplified to $R\theta$. To apply these rotations to a point (x, y), calculate:

$$R_{90°}(x, y) = (-y, x)$$

$$R_{180°}(x, y) = (-x, -y)$$

$$R_{270°}(x, y) = (y, -x)$$

For polygons defined by vertices, apply the rotation to each vertex and connect the resulting points to form the image. For example, to rotate triangle ABC with vertices $A(1, 2)$, $B(3, 2)$, and $C(2, 5)$ by 90° counterclockwise around the origin:

$$A(1, 2) \rightarrow A'(-2, 1)$$

$$B(3, 2) \rightarrow B'(-2, 3)$$

$$C(2, 5) \rightarrow C'(-5, 2)$$

Connecting these rotated vertices creates triangle A'B'C', the image of triangle ABC under the 90° counterclockwise rotation.

Rotations have several key properties:

1. The pre-image and image are congruent figures.
2. Corresponding angles remain equal.
3. Corresponding segments remain equal in length.
4. The orientation of the figure is preserved.
5. A rotation by 360° maps each point back to itself.
6. Two successive rotations around the same center result in a single rotation around that center by the sum of the angles.

In geometric constructions, rotations can be performed using compass and straightedge by transferring angles. In real-world applications, rotations model circular movement, such as the motion of gears, wheels, and celestial bodies.

Dilations

A dilation is a transformation that resizes a figure without changing its shape. Each point of the original figure (pre-image) moves along a ray emanating from a fixed point called the center of dilation. The distance from each point to the center of dilation is multiplied by a constant factor called the scale factor. Dilations preserve angle measure but not distance, making them non-rigid transformations.

In coordinate geometry, dilations are typically performed with the origin (0, 0) as the center. The mapping rule for a dilation with center at the origin and scale factor k is:

$$(x, y) \rightarrow (kx, ky)$$

The value of k determines the size of the image relative to the pre-image. If $k > 1$, the image is an enlargement of the pre-image; if $0 < k < 1$, the image is a reduction; if $k = 1$, the image is congruent to the pre-image; if k < 0, the image is both resized and reflected through the center of dilation.

For example, under the dilation with center at the origin and scale factor $k = 2$, the point (3, 4) maps to $(2 \times 3, 2 \times 4) = (6, 8)$. Every point's distance from the origin is doubled.

For a dilation with center at an arbitrary point (h, k) and scale factor s, the mapping rule becomes:

$$(x, y) \rightarrow \big(h + s(x - h), k + s(y - k)\big)$$

This formula first shifts the center of dilation to the origin, applies the dilation, and then shifts back.

The notation $D(C, k)$ is often used to represent a dilation with center C and scale factor k. For dilations with center at the origin, this can be simplified to Dk. To apply a dilation with center at the origin and scale factor k to a point (x, y), calculate:

$$D_k(x, y) = (kx, ky)$$

For polygons defined by vertices, apply the dilation to each vertex and connect the resulting points to form the image. For example, to dilate triangle ABC with vertices $A(1, 2)$, $B(3, 2)$, and $C(2, 5)$ by a scale factor of 3 from the origin:

$$A(1, 2) \rightarrow A'(3 \times 1, 3 \times 2) = A'(3, 6)$$

$$B(3, 2) \rightarrow B'(3 \times 3, 3 \times 2) = B'(9, 6)$$

$$C(2,5) \rightarrow C'(3 \times 2, 3 \times 5) = C'(6,15)$$

Connecting these dilated vertices creates triangle A'B'C', the image of triangle ABC under the dilation with scale factor 3.

Dilations have several key properties:

1. The pre-image and image are similar figures.
2. Corresponding angles remain equal.
3. The ratio of corresponding segment lengths equals the scale factor.
4. The ratio of the area of the image to the area of the pre-image equals the square of the scale factor.
5. The ratio of the perimeter of the image to the perimeter of the pre-image equals the scale factor.
6. A dilation with scale factor -1 is equivalent to a 180° rotation around the center of dilation.

In geometric constructions, dilations can be performed using the properties of similar triangles. In real-world applications, dilations model scaling, such as in maps, blueprints, and photography.

Symmetry

Symmetry occurs when a figure remains invariant (unchanged) under certain transformations. The two primary types of symmetry in two-dimensional geometry are reflectional symmetry and rotational symmetry.

Reflectional symmetry, also called line symmetry or mirror symmetry, occurs when a figure can be mapped onto itself by a reflection over a line called the axis of symmetry. A figure may have multiple axes of symmetry. For example, an equilateral triangle has three axes of symmetry, a square has four, and a circle has infinitely many.

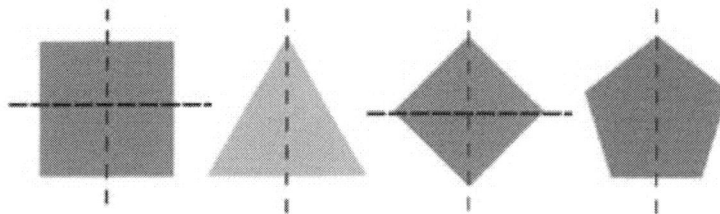

Reflectional symmetry

To determine whether a figure has reflectional symmetry, one can:

1. Check if a line divides the figure into two halves that are reflections of each other
2. Fold the figure along a potential axis of symmetry to see if both halves coincide.
3. In coordinate geometry, test whether the figure remains invariant under reflection over specific lines.

For example, the parabola with equation $y = x^2$ has reflectional symmetry over the y-axis because for every point (a, b) on the parabola, the point $(-a, b)$ is also on the parabola.

Rotational symmetry occurs when a figure can be mapped onto itself by a rotation around a point called the center of rotational symmetry. The angle of the smallest such rotation (other than 0°) is called the angle of rotational symmetry, and the number of distinct positions the figure occupies during a complete 360° rotation is called the order of rotational symmetry.

For example, an equilateral triangle has rotational symmetry of order 3 because it can be rotated by 120° to map onto itself, and it occupies three distinct positions in a complete 360° rotation. A square has rotational symmetry of order 4 with an angle of rotational symmetry of 90°.

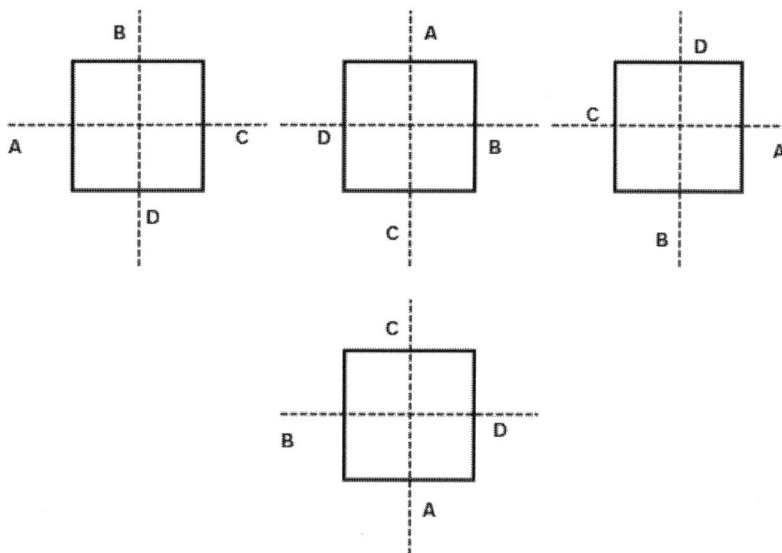

Rotational symmetry of Square at 90°

To determine whether a figure has rotational symmetry, one can:

1. Identify a potential center of rotational symmetry (often the center of the figure).
2. Rotate the figure around this center and check if it maps onto itself for any angle less than 360°.
3. In coordinate geometry, test whether the figure remains invariant under rotation around specific points.

For example, the circle with equation $x^2 + y^2 = r^2$ has rotational symmetry of infinite order around the origin because it remains invariant under any rotation around the origin.

Symmetry has applications in various fields:

1. In art and design, symmetry creates balance and aesthetic appeal.
2. In nature, many organisms and structures exhibit symmetry, such as flowers, snowflakes, and crystal formations.
3. In physics, symmetry principles underlie fundamental laws and conservation theorems.
4. In architecture, symmetry often guides the design of buildings and structures.

Understanding symmetry also simplifies problem-solving in geometry. For instance, if a figure has reflectional symmetry, calculations need only be performed on one side of the axis of symmetry and then mapped to the other side.

Composition of Transformations

A composition of transformations is the application of two or more transformations in sequence. The notation for composing transformations uses function composition: if S and T are transformations, then S ∘ T denotes the transformation that results from first applying T and then applying S.

Compositions of transformations can often be simplified to a single transformation. Understanding these simplifications helps in analyzing complex geometric situations and solving problems efficiently.

The composition of two translations results in a single translation. If $T(h_1, k_1)$ and $T(h_2, k_2)$ are translations, then:

$$T_{(h_1,k_1)} \circ T_{(h_2,k_2)} = T_{(h_1+h_2,k_1+k_2)}$$

For example, applying translation $T(3,4)$ followed by translation $T(2,-5)$ is equivalent to a single translation $T(3+2, 4+(-5)) = T(5,-1)$.

The composition of two reflections depends on the relationship between the lines of reflection:

1. If the lines of reflection are parallel, the composition is equivalent to a translation perpendicular to the lines of reflection, with magnitude twice the distance between the lines.
2. If the lines of reflection intersect, the composition is equivalent to a rotation around the point of intersection, with angle twice the measure of the acute angle between the lines.

For example, a reflection over the x-axis followed by a reflection over the y-axis is equivalent to a 180° rotation around the origin:

$$r_{y-axis} \circ r_{x-axis}(x, y) = r_{y-axis}(x, -y) = (-x, -y) = R_{180°}(x, y)$$

The composition of two rotations around the same center results in a single rotation around that center, with angle equal to the sum of the angles of the component rotations. If $R(O, \theta_1)$ and $R(O, \theta_2)$ are rotations around point O, then:

$$R_{(O,\theta_1)} \circ R_{(O,\theta_2)} = R_{(O,\theta_1+\theta_2)}$$

For example, a 60° rotation followed by a 75° rotation around the same center is equivalent to a 135° rotation around that center.

The composition of two dilations with the same center results in a single dilation with that center, with scale factor equal to the product of the component scale factors. If $D(C, k_1)$ and $D(C, k_2)$ are dilations with center C, then:

$$D_{(C,k_1)} \circ D_{(C,k_2)} = D_{(C,k_1 \times k_2)}$$

For example, a dilation with scale factor 2 followed by a dilation with scale factor 3, both centered at the origin, is equivalent to a single dilation with scale factor $2 \times 3 = 6$ centered at the origin.

Compositions of different types of transformations can also be analyzed:

1. A translation followed by a reflection is equivalent to a reflection over a parallel line.

2. A rotation followed by a translation can be expressed as a single rotation around a different center.

3. A dilation followed by a translation cannot generally be simplified to a single transformation.

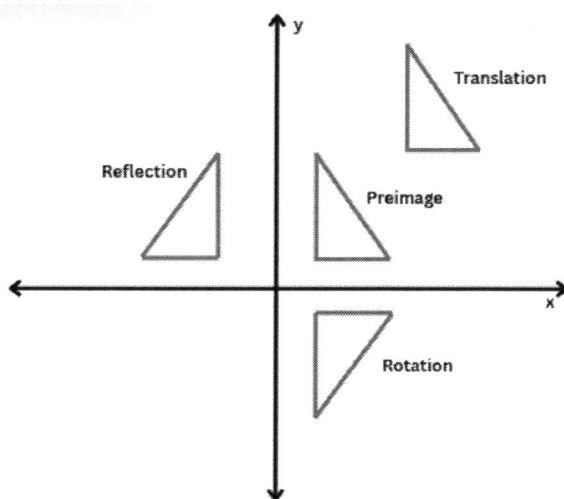

In coordinate geometry, compositions of transformations can be analyzed algebraically by applying the transformation formulas in sequence. For example, to find the image of point $(3, 2)$ under the composition of a $90°$ rotation around the origin followed by a translation $T(4, -1)$:

1. Apply the $90°$ rotation: $R_{90}°(3, 2) = (-2, 3)$

2. Apply the translation: $T(4, -1)(-2, 3) = (-2 + 4, 3 + (-1)) = (2, 2)$

The final image of point $(3, 2)$ under this composition is $(2, 2)$.

Practice Problems

1. Find the coordinates of triangle ABC with vertices $A(1, 3)$, $B(4, 2)$, and $C(2, 5)$ after translation $T(3, -4)$.

2. Determine the coordinates of quadrilateral PQRS with vertices $P(3, 1)$, $Q(5, 2)$, $R(4, 4)$, and $S(2, 3)$ after reflection over the x-axis.

3. Find the image of point $(3, -2)$ under a $90°$ counterclockwise rotation around the origin, followed by a reflection over the line $y = x$.

4. Triangle DEF has vertices $D(-2, 1)$, $E(0, 3)$, and $F(2, 0)$. Find the coordinates of the vertices of the image triangle after a dilation with center at the origin and scale factor 3.

5. Determine whether the following figure has reflectional symmetry, rotational symmetry, both, or neither. If it has reflectional symmetry, identify all axes of symmetry. If it has rotational symmetry, identify the center and order of rotational symmetry: a regular hexagon.

Solutions

1. To find the image of triangle ABC under translation T(3,-4), apply the transformation to each vertex:

$$A(1,3) \rightarrow A'(1 + 3, 3 + (-4)) = A'(4, -1)$$

$$B(4,2) \rightarrow B'(4 + 3, 2 + (-4)) = B'(7, -2)$$

$$C(2,5) \rightarrow C'(2 + 3, 5 + (-4)) = C'(5, 1)$$

The image triangle A'B'C' has vertices at $(4, -1)$, $(7, -2)$, and $(5, 1)$.

2. To find the image of quadrilateral PQRS under reflection over the x-axis, apply the transformation to each vertex:

$$P(3,1) \rightarrow P'(3, -1)$$

$$Q(5,2) \rightarrow Q'(5, -2)$$

$$R(4,4) \rightarrow R'(4, -4)$$

$$S(2,3) \rightarrow S'(2, -3)$$

The image quadrilateral P'Q'R'S' has vertices at $(3, -1)$, $(5, -2)$, $(4, -4)$, and $(2, -3)$.

3. To find the image of point $(3, -2)$ under a 90° counterclockwise rotation followed by a reflection over $y = x$:

Step 1: Apply the 90° rotation: $R_{90°}(3, -2) = (-(-2), 3) = (2,3)$

Step 2: Apply the reflection over $y = x$: $r_{y=x}(2,3) = (3,2)$

The final image of point (3, -2) under this composition is (3, 2).

4. To find the image of triangle DEF under a dilation with center at the origin and scale factor 3:

$$D(-2,1) \rightarrow D'(3 \times (-2), 3 \times 1) = D'(-6, 3)$$

$$E(0,3) \rightarrow E'(3 \times 0, 3 \times 3) = E'(0, 9)$$

$$F(2,0) \rightarrow F'(3 \times 2, 3 \times 0) = F'(6, 0)$$

The image triangle D'E'F' has vertices at (-6, 3), (0, 9), and (6, 0).

5. A regular hexagon has both reflectional and rotational symmetry:

   a. Reflectional symmetry: A regular hexagon has 6 axes of symmetry:

   Three axes pass through opposite vertices

   Three axes pass through the midpoints of opposite sides

b. Rotational symmetry: A regular hexagon has rotational symmetry of order 6 around its center. It can be rotated by multiples of 60° (360° ÷ 6 = 60°) and map onto itself.

CONGRUENCE AND SIMILARITY

Triangle Classification and Properties

Triangles form the foundation of many geometric concepts and are classified according to their side lengths and angle measures.

Based on side lengths, triangles are classified into three categories:

1. Equilateral triangles have three congruent sides. If ABC is an equilateral triangle, then AB = BC = AC. An equilateral triangle also has three congruent angles, each measuring 60°.

2. Isosceles triangles have at least two congruent sides. If ABC is an isosceles triangle with AB = AC, then B and C are called the base angles, and these angles are congruent. The side BC, opposite the vertex angle A, is called the base. The altitude from A to BC is also the perpendicular bisector of BC and the angle bisector of angle A.

3. Scalene triangles have no congruent sides. In a scalene triangle, all sides have different lengths, and all angles have different measures.

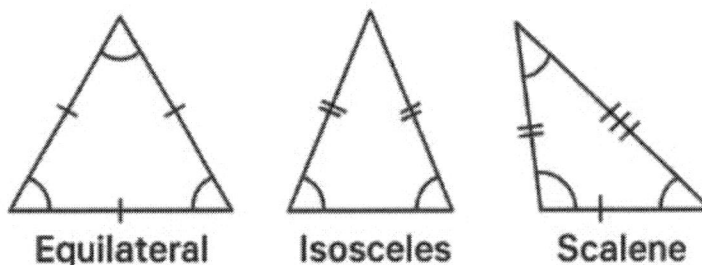

Equilateral   Isosceles   Scalene

Based on angle measures, triangles are classified into three categories:

1. Acute triangles have all three angles measuring less than 90°.

2. Right triangles have one angle measuring exactly 90°. The side opposite the right angle is called the hypotenuse, and the other two sides are called legs.

   The Pythagorean theorem states that in a right triangle, the square of the length of the hypotenuse equals the sum of the squares of the lengths of the legs:

$$a^2 + b^2 = c^2$$

   where c is the length of the hypotenuse, and a and b are the lengths of the legs.

3. Obtuse triangles have one angle measuring more than 90°.

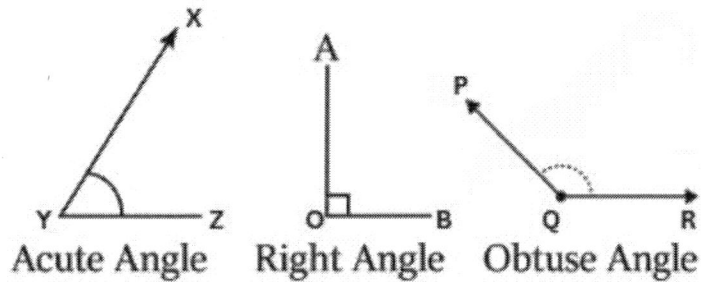

Acute Angle   Right Angle   Obtuse Angle

Every triangle satisfies several fundamental properties:

1. The sum of the interior angles of a triangle equals 180°:

$$m\angle A + m\angle B + m\angle C = 180°$$

2. The exterior angle of a triangle equals the sum of the two non-adjacent interior angles:

$$m\angle ACD = m\angle A + m\angle B$$

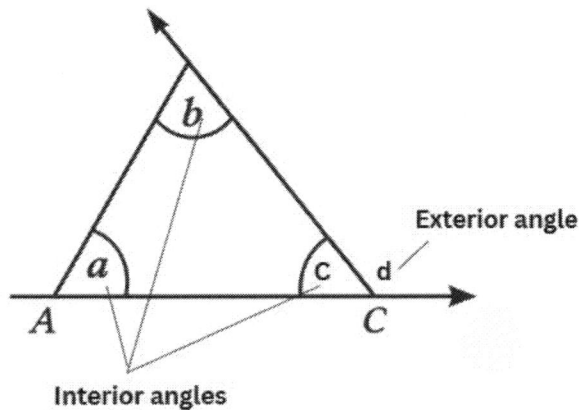

3. The sum of the lengths of any two sides of a triangle is greater than the length of the third side (Triangle Inequality Theorem):

$$a + b > c$$

$$b + c > a$$

$$a + c > b$$

where a, b, and c are the lengths of the sides of the triangle.

4. The medians of a triangle are segments connecting each vertex to the midpoint of the opposite side. The three medians of a triangle intersect at a single point called the centroid, which divides each median in a 2:1 ratio (the portion from the vertex is twice as long as the portion from the centroid to the midpoint).

5. The perpendicular bisectors of the sides of a triangle intersect at a single point called the circumcenter, which is equidistant from the three vertices of the triangle. This point is the center of the circumscribed circle of the triangle.

6. The angle bisectors of a triangle intersect at a single point called the incenter, which is equidistant from the three sides of the triangle. This point is the center of the inscribed circle of the triangle.

7. The altitudes of a triangle are perpendicular segments from each vertex to the opposite side (or its extension). The three altitudes of a triangle intersect at a single point called the orthocenter.

Properties of Congruent Figures

Two geometric figures are congruent if they have exactly the same size and shape. Formally, two figures are congruent if one can be mapped onto the other by a sequence of rigid transformations (translations, rotations, and reflections). Congruent figures have corresponding parts (angles and sides) that are congruent.

For congruent triangles, the corresponding sides have equal lengths, and the corresponding angles have equal measures. If triangles ABC and DEF are congruent, written as $\triangle ABC \cong \triangle DEF$, then:

$$AB = DE$$

$$BC = EF$$

$$AC = DF$$

$$m\angle A = m\angle D$$

$$m\angle B = m\angle E$$

$$m\angle C = m\angle F$$

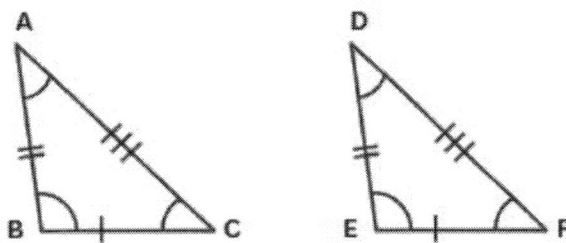

Congruent triangles

The concept of congruence extends to other polygons as well. Two polygons are congruent if they have the same number of sides, and corresponding sides and angles are congruent. This means that congruent polygons have equal perimeters and equal areas.

Congruence preserves several geometric properties:

1. Length: Corresponding segments in congruent figures have equal lengths.
2. Angle measure: Corresponding angles in congruent figures have equal measures.
3. Area: Congruent figures have equal areas.
4. Perimeter: Congruent figures have equal perimeters.

In coordinate geometry, two figures are congruent if one can be mapped onto the other by a sequence of rigid transformations, which in the coordinate plane correspond to translations, rotations, and reflections. The distance between any two points in one figure equals the distance between the corresponding points in the congruent figure.

For example, triangles with vertices at $(0, 0)$, $(3, 0)$, $(0, 4)$ and $(5, 7)$, $(8, 7)$, $(5, 11)$ are congruent because the second triangle can be obtained from the first by a translation of 5 units right and 7 units up. The distance between any two vertices in the first triangle equals the distance between the corresponding vertices in the second triangle.

Proving Congruence (SSS, SAS, ASA)

Several criteria exist for proving that two triangles are congruent without having to verify that all corresponding parts are congruent. These criteria provide efficient methods for establishing triangle congruence based on limited information.

The Side-Side-Side (SSS) congruence criterion states that if the three sides of one triangle are congruent to the three sides of another triangle, then the triangles are congruent. That is, if $AB = DE$, $BC = EF$, and $AC = DF$, then $\triangle ABC \cong \triangle DEF$.

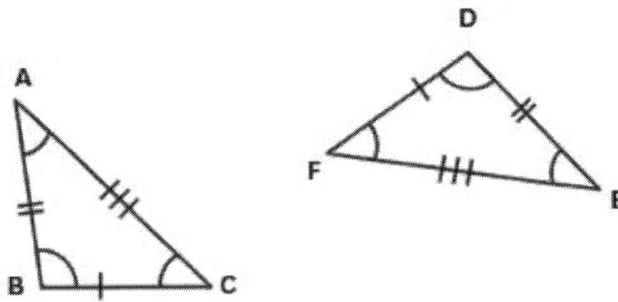

SSS congruence

The Side-Angle-Side (SAS) congruence criterion states that if two sides and the included angle of one triangle are congruent to two sides and the included angle of another triangle, then the triangles are congruent. That is, if $AB = DE, m\angle B = m\angle E$, and $BC = EF$, then $\triangle ABC \cong \triangle DEF$.

The Angle-Side-Angle (ASA) congruence criterion states that if two angles and the included side of one triangle are congruent to two angles and the included side of another triangle, then the triangles are congruent. That is, if $m\angle A = m\angle D, AB = DE$, and $m\angle B = m\angle E$, then $\triangle ABC \cong \triangle DEF$.

The Angle-Angle-Side (AAS) congruence criterion states that if two angles and a non-included side of one triangle are congruent to two angles and the corresponding non-included side of another triangle, then the triangles are congruent. That is, if $m\angle A = m\angle D, m\angle B = m\angle E$, and $BC = EF$, then $\triangle ABC \cong \triangle DEF$.

It's important to note that "Angle-Angle-Angle" (AAA) is not a valid congruence criterion. If three angles of one triangle are congruent to three angles of another triangle, the triangles are similar but not necessarily congruent. Similarly, "Side-Side-Angle" (SSA) is not a valid congruence criterion in general, though it works in specific cases, such as when the angle is a right angle (Hypotenuse-Leg or HL congruence for right triangles).

These congruence criteria can be proven using the rigid transformation definition of congruence. For example, the SSS criterion can be demonstrated by showing that two triangles with congruent corresponding sides can be mapped onto each other by a sequence of rigid transformations.

The congruence criteria are powerful tools for establishing the congruence of triangles when limited information is available, which is particularly useful in proofs and problem solving.

Similar Figures

Two geometric figures are similar if they have the same shape but not necessarily the same size. Formally, two figures are similar if one can be mapped onto the other by a sequence of rigid transformations (translations, rotations, and reflections) followed by a dilation. Similar figures have corresponding angles that are congruent and corresponding sides that are proportional.

For similar triangles, if triangles ABC and DEF are similar, written as $\triangle ABC \sim \triangle DEF$, then:

$$m\angle A = m\angle D$$

$$m\angle B = m\angle E$$

$$m\angle C = m\angle F$$

and

$$\frac{AB}{DE} = \frac{BC}{EF} = \frac{AC}{DF} = r$$

where r is the ratio of similarity (or scale factor).

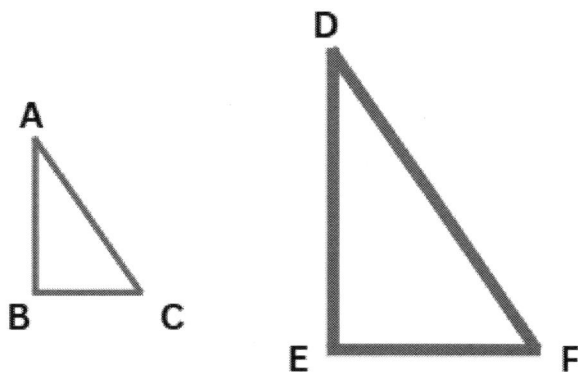

Similar triangles

Several criteria exist for proving that two triangles are similar:

The Angle-Angle (AA) similarity criterion states that if two angles of one triangle are congruent to two angles of another triangle, then the triangles are similar. Since the sum of the angles in any triangle is 180°, if two angles are congruent, the third angle must also be congruent.

40

The Side-Side-Side (SSS) similarity criterion states that if the three sides of one triangle are proportional to the three sides of another triangle, then the triangles are similar. That is, if $\frac{AB}{DE} = \frac{BC}{EF} = \frac{AC}{DF}$, then $\triangle ABC \sim \triangle DEF$.

The Side-Angle-Side (SAS) similarity criterion states that if two sides of one triangle are proportional to two sides of another triangle, and the included angles are congruent, then the triangles are similar. That is, if $\frac{AB}{DE} = \frac{BC}{EF}$ and $m\angle B = m\angle E$, then $\triangle ABC \sim \triangle DEF$.

Similarity extends to other polygons as well. Two polygons are similar if they have the same number of sides, corresponding angles are congruent, and corresponding sides are proportional. This means that similar polygons have the same shape but may have different sizes.

Similarity preserves several geometric properties and yields important relationships:

1. Angle measure: Corresponding angles in similar figures are congruent.
2. Length ratio: The ratio of corresponding sides in similar figures is constant.
3. Area ratio: The ratio of the areas of similar figures equals the square of the ratio of corresponding sides: $\frac{Area_1}{Area_2} = r^2$.
4. Perimeter ratio: The ratio of the perimeters of similar figures equals the ratio of corresponding sides: $\frac{Perimeter_1}{Perimeter_2} = r$.

In coordinate geometry, two figures are similar if one can be mapped onto the other by a sequence of rigid transformations followed by a dilation. The distance between any two points in one figure is proportional to the distance between the corresponding points in the similar figure, with the proportion being the scale factor.

Properties of Similar Triangles

Similar triangles possess several special properties that properties extend beyond the basic similarity criteria and provide insights into the relationships between corresponding parts of similar triangles.

One fundamental property of similar triangles is that the ratio of the areas of two similar triangles equals the square of the ratio of their corresponding sides. If triangles ABC and DEF are similar with a similarity ratio of r, then:

$$\frac{Area\ of\ \triangle ABC}{Area\ of\ \triangle DEF} = r^2$$

This relationship arises because area is a two-dimensional measure, so it scales by the square of the linear scaling factor.

Similarly, the ratio of the perimeters of two similar triangles equals the ratio of their corresponding sides. If triangles ABC and DEF are similar with a similarity ratio of r, then:

$$\frac{\textbf{\textit{Perimeter of }} \triangle \textbf{\textit{ABC}}}{\textbf{\textit{Perimeter of }} \triangle \textbf{\textit{DEF}}} = r$$

This property holds because perimeter is a one-dimensional measure, so it scales linearly with the scaling factor.

Another important property involves the altitudes of similar triangles. If triangles ABC and DEF are similar with a similarity ratio of r, and if h_1 and h_2 are corresponding altitudes, then:

$$\frac{h_1}{h_2} = r$$

The same proportion applies to medians, angle bisectors, and other corresponding segments in similar triangles.

The midsegment theorem is a special case of triangle similarity. A midsegment of a triangle is a segment connecting the midpoints of two sides of the triangle. The midsegment theorem states that the midsegment is parallel to the third side of the triangle and has a length equal to half the length of the third side.

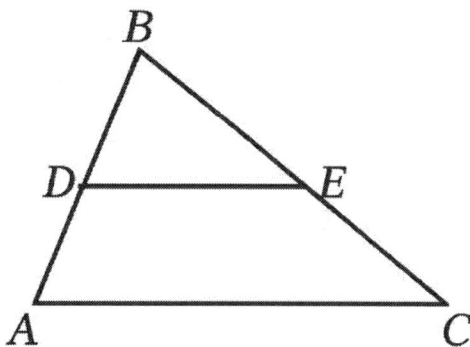

If D is the midpoint of AB and E is the midpoint of BC, then,
1. DE is one-half of AC.
2. DE is parallel to AC.

Midsegment theorem

This theorem can be proven using similar triangles. If M is the midpoint of AB and N is the midpoint of AC in triangle ABC, then triangles AMN and ABC are similar with a similarity ratio of 1:2, which implies that MN is parallel to BC and MN = ½BC.

42

The triangle proportionality theorem (or basic proportionality theorem) states that if a line parallel to one side of a triangle intersects the other two sides, then it divides those sides proportionally. Conversely, if a line divides two sides of a triangle proportionally, then it is parallel to the third side.

This theorem is a direct application of similar triangles. If line DE is parallel to side BC of triangle ABC, with D on AB and E on AC, then:

$$\frac{AD}{DB} = \frac{AE}{EC}$$

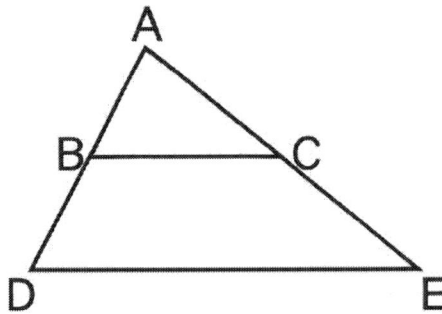

If BC is parallel to DE, then

$$\frac{AB}{BD} = \frac{AC}{CE}$$

Another application of similar triangles is in solving problems involving shadows and indirect measurements. By using the properties of similar triangles, one can determine the height of tall objects like trees or buildings by measuring shadows or using mirrors.

Applications of Similarity

One of the most common applications of similarity is in indirect measurement. When direct measurement is impractical, similar triangles can be used to determine unknown dimensions. For example, to find the height of a tree, one can use the shadow method: measure the tree's shadow length, the height of a known object (like a person), and the object's shadow length. The proportion:

$$\frac{Tree\ height}{Object\ height} = \frac{Tree\ shadow}{Object\ shadow}$$

allows for calculation of the tree's height.

Another application involves scale models and maps. Architectural models, engineering prototypes, and geographical maps all use similarity principles. The scale factor establishes the relationship between the model and the actual object. If a map has a scale of $1:10{,}000$, then 1 cm on the map represents 10,000 cm (or 100 m) in the real world. Areas on the map scale by the square of this factor, so 1 square centimeter represents 100,000,000 *square centimeters* (or 10,000 square meters) in reality.

In art and design, similarity principles govern perspective drawing. Parallel lines in three-dimensional space appear to converge at vanishing points in two-dimensional perspective drawings, creating similar triangles that help establish correct proportions.

The concept of similarity also applies to fractal geometry, where self-similar patterns repeat at different scales. Fractals like the Sierpinski triangle exhibit the same pattern regardless of how much one zooms in, demonstrating perfect self-similarity.

In architecture, similar triangles help solve problems related to stability and structural design. For instance, similar triangles are used to analyze truss systems, where forces are distributed through triangular structures.

Photography employs similarity principles in lens equations. The relationship between object distance (p), image distance (q), and focal length (f) follows:

$$\frac{1}{p} + \frac{1}{q} = \frac{1}{f}$$

This equation arises from similar triangles formed by light rays passing through the lens.

In computer graphics, similarity transformations are fundamental to scaling objects while preserving their shape. These transformations maintain the proportions of the original object while changing its size.

Navigation and surveying rely heavily on similarity principles. Triangulation, a technique used to determine distances and positions, employs similar triangles to calculate unknown measurements based on known angles and distances.

The golden ratio (approximately 1.618), found in various natural patterns and considered aesthetically pleasing, creates similar triangles when applied to geometric constructions. The golden rectangle, for instance, can be subdivided into a square and another golden rectangle, creating similar rectangles.

These applications demonstrate the versatility and practical importance of similarity concepts in solving real-world problems across diverse fields.

Practice Problems

1. In triangle ABC, D is a point on side AB such that $AD = (1/3)AB$. Point E is on side AC such that $AE = (1/3)AC$. Prove that DE is parallel to BC and find the ratio of the area of triangle ADE to the area of triangle ABC.

2. Triangle ABC has vertices $A(2,3)$, $B(8,3)$, and $C(4,7)$. Determine whether triangle ABC is isosceles, and if so, identify the congruent sides.

3. Triangles ABC and DEF have sides with the following measurements: $AB = 6\,cm$, $BC = 8\,cm$, $AC = 10\,cm$, $DE = 9\,cm$, $EF = 12\,cm$, and $DF = 15\,cm$. Are these triangles similar? If so, find the similarity ratio and determine which sides correspond.

4. In quadrilateral $ABCD$, the diagonals AC and BD intersect at point P. If $AP = 3PC$ and $BP = 2PD$, determine whether triangles ABP and CDP are similar. If they are similar, find the similarity ratio.

Solutions

1. To prove that DE is parallel to BC and find the area ratio:

Given: In triangle ABC, D is on AB with $AD = (1/3)AB$, and E is on AC with $AE = (1/3)AC$.

First, note that: $AD = (1/3)AB$ implies $BD = \left(\frac{2}{3}\right)AB$, $AE = (1/3)AC$ implies $EC = (2/3)AC$

By the triangle proportionality theorem, if a line divides two sides of a triangle proportionally, then it is parallel to the third side. Since D divides AB in the ratio 1:2 and E divides AC in the same ratio $1:2$, DE is parallel to BC.

For the area ratio, when a line parallel to one side of a triangle forms a similar triangle, the ratio of the areas equals the square of the ratio of the corresponding sides. Since $DE \parallel BC$ and the ratio of $AD:AB = AE:AC = 1:3$, the similarity ratio is $1:3$.

Therefore:

$$\frac{Area\ of\ \triangle ABC}{Area\ of\ \triangle DEF} = (\frac{1}{3})^2$$

The area of triangle ADE is 1/9 the area of triangle ABC.

2. To determine if triangle ABC with vertices $A(2,3)$, $B(8,3)$, and $C(4,7)$ is isosceles:

Calculate the lengths of the three sides:

$$AB = \sqrt{[(8-2)^2 + (3-3)^2]} = \sqrt{[36+0]} = 6\ units$$

$$BC = \sqrt{[(4-8)^2 + (7-3)^2]} - \sqrt{[16+16]} = \sqrt{32} = 4\sqrt{2}\ units$$

$$AC = \sqrt{[(4-2)^2 + (7-3)^2]} = \sqrt{[4+16]} = \sqrt{20} = 2\sqrt{5}\ units$$

Since $AB \neq BC$ and $AB \neq AC$ and $BC \neq AC$, triangle ABC is not isosceles. It is a scalene triangle.

45

3. To determine if triangles ABC and DEF with sides $AB = 6\ cm$, $BC = 8\ cm$, $AC = 10\ cm$, $DE = 9\ cm$, $EF = 12\ cm$, and $DF = 15\ cm$ are similar:

Check if corresponding sides are proportional (SSS similarity criterion):

$$\frac{DE}{AB} = \frac{9}{6} = \frac{3}{2}$$

$$\frac{EF}{BC} = \frac{12}{8} = \frac{3}{2}$$

$$\frac{DF}{AC} = \frac{15}{10} = \frac{3}{2}$$

Since all three ratios are equal (3/2), the triangles are similar by the SSS similarity criterion. The similarity ratio is 3:2.

The corresponding sides are: DE corresponds to AB, EF corresponds to BC, DF corresponds to AC

4. To determine if triangles ABP and CDP are similar:

Given: In quadrilateral ABCD, diagonals AC and BD intersect at P, with $AP = 3PC$ and $BP = 2PD$.

The proportion $AP:PC = 3:1$ means $AP = 3/(3+1) = 3/4$ of AC, and $PC = 1/4$ of AC. The proportion $BP:PD = 2:1$ means $BP = 2/(2+1) = 2/3$ of BD, and $PD = 1/3$ of BD.

For triangles ABP and CDP to be similar, we need to check if two angles are congruent or if corresponding sides are proportional.

Angle APB in triangle ABP corresponds to angle CPD in triangle CDP.

Angle BAP in triangle ABP corresponds to angle DCP in triangle CDP. These angles are vertical angles formed by the intersection of AC and BD, so they are congruent.

Similarly, angle ABP in triangle ABP corresponds to angle CDP in triangle CDP. These are also vertical angles, so they are congruent.

Since two angles of triangle ABP are congruent to two angles of triangle CDP, the triangles are similar by the AA similarity criterion.

The similarity ratio equals the ratio of corresponding sides: $\frac{AB}{CD} = \frac{AP}{CP} = \frac{3PC}{PC} = 3:1$

Therefore, triangles ABP and CDP are similar with a similarity ratio of 3:1.

POLYGONS AND QUADRILATERALS

Properties of Polygons

A polygon is a closed plane figure formed by connecting line segments, where each line segment intersects exactly two others at its endpoints. The line segments, called sides, form the boundary of the polygon, and the points where the sides meet are called vertices. Polygons are classified by the number of sides: triangles (3 sides), quadrilaterals (4 sides), pentagons (5 sides), hexagons (6 sides), and so forth.

A polygon is convex if no line containing a side of the polygon contains points in the interior of the polygon. Equivalently, a polygon is convex if all interior angles are less than 180°. If at least one interior angle exceeds 180°, the polygon is concave.

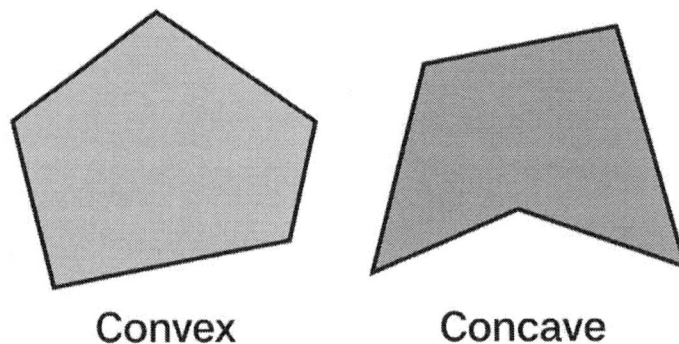

Convex   Concave

Convex and concave polygons

A regular polygon has all sides congruent and all interior angles congruent. Examples include equilateral triangles, squares, regular pentagons, and regular hexagons. The measure of each interior angle in a regular n-sided polygon is given by:

$$m\angle = \frac{(n-2) \times 180°}{n}$$

For instance, each interior angle in a regular pentagon (n = 5) measures:

$$m\angle = \frac{(5-2) \times 180°}{5} = \frac{540°}{5} = 108°$$

The sum of the interior angles of any n-sided polygon equals $(n - 2) \times 180°$. This formula can be derived by dividing the polygon into $(n - 2)$ triangles, each containing $180°$. For example, the sum of interior angles in a hexagon $(n = 6)$ is:

$$(6 - 2) \times 180° = 4 \times 180° = 720°$$

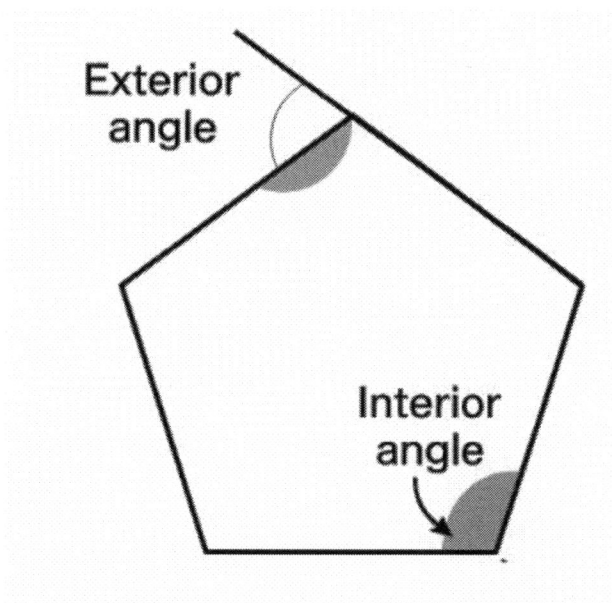

A Polygon

The exterior angle of a polygon at a vertex is the angle formed by extending one side and measuring the angle between this extension and the adjacent side. The sum of the exterior angles, one at each vertex of a polygon, always equals $360°$, regardless of the number of sides. The measure of each exterior angle in a regular n-sided polygon is:

$$m\angle_{ext} = \frac{360°}{n}$$

For example, each exterior angle in a regular octagon $(n = 8)$ measures:

$$m\angle_{ext} = \frac{360°}{8} = 45°$$

The perimeter of a polygon is the sum of the lengths of all its sides. For a regular n-sided polygon with side length s, the perimeter is:

$$P = n \times s$$

The area of a regular n-sided polygon with side length s can be calculated using the apothem a (the distance from the center to the midpoint of any side):

$$A = \frac{1}{2} \times P \times a = \frac{1}{2} \times n \times s \times a$$

The apothem can be found using the formula:

$$a = \frac{s}{2tan\left(\frac{180°}{n}\right)}$$

Angles of Polygons

The measures of interior and exterior angles follow patterns that provide insight into the structure of polygons.

The interior angle sum formula states that the sum of all interior angles in an n-sided polygon equals (n - 2) × 180°:

$$S = (n - 2) \times 180°$$

This formula can be derived by dividing the polygon into (n - 2) triangles by drawing all diagonals from one vertex. Since each triangle has an angle sum of 180°, the entire polygon has an angle sum of $(n - 2) \times 180°$.

For a regular polygon, where all interior angles have the same measure, each interior angle equals:

$$m\angle = \frac{(n - 2) \times 180°}{n}$$

As the number of sides increases, the measure of each interior angle in a regular polygon approaches 180°. This reflects the fact that a circle can be viewed as a regular polygon with infinitely many sides, where each interior angle is 180°.

The exterior angle of a polygon at a vertex is supplementary to the interior angle at that vertex. Thus:

$$m\angle_{ext} + m\angle_{int} = 180°$$

The sum of all exterior angles in any polygon, taking one exterior angle at each vertex, always equals 360°. This is true regardless of the number of sides or whether the polygon is regular. For a regular polygon with n sides, each exterior angle measures:

$$m\angle_{ext} = \frac{360°}{n}$$

The exterior angle theorem for triangles—stating that an exterior angle equals the sum of the two non-adjacent interior angles—extends to polygons. An exterior angle of a polygon equals the sum of all interior angles except the one adjacent to it.

When a polygon is inscribed in a circle (all vertices lie on the circle), special angle relationships emerge. In a cyclic quadrilateral, opposite angles are supplementary, meaning they sum to 180°:

$$m\angle A + m\angle C = m\angle B + m\angle D = 180°$$

When a polygon is circumscribed around a circle (all sides are tangent to the circle), the segments from the center of the circle to the points of tangency are perpendicular to the respective sides.

Angle bisectors in polygons have interesting properties. In triangles, the angle bisectors are concurrent at the incenter. In regular polygons, all angle bisectors meet at the center of the polygon, which is also the center of the inscribed and circumscribed circles.

In coordinate geometry, angles between lines can be calculated using the inverse tangent function and slopes. If two lines have slopes m_1 and m_2, the angle θ between them is given by:

$$tan\theta = \left| \frac{m_2 - m_1}{1 + m_1 m_2} \right|$$

This formula allows for precise calculation of angles in polygons placed in a coordinate system.

Properties of Parallelograms

A parallelogram is a quadrilateral with opposite sides parallel. This fundamental definition leads to several important properties that characterize parallelograms and distinguish them from other quadrilaterals.

The defining property, opposite sides being parallel, can be written as:

$$AB \parallel DC \ and \ AD \parallel BC$$

From this definition, several key properties can be proven:

Opposite sides of a parallelogram are congruent:

$$AB = DC \ and \ AD = BC$$

Opposite angles of a parallelogram are congruent:

$$m\angle A = m\angle C \text{ and } m\angle B = m\angle D$$

Consecutive angles of a parallelogram are supplementary (sum to 180°):

$$m\angle A + m\angle B = m\angle B + m\angle C = m\angle C + m\angle D = m\angle D + m\angle A = 180°$$

The diagonals of a parallelogram bisect each other:

$$AO = OC \text{ and } BO = OD$$

where O is the point of intersection of the diagonals.

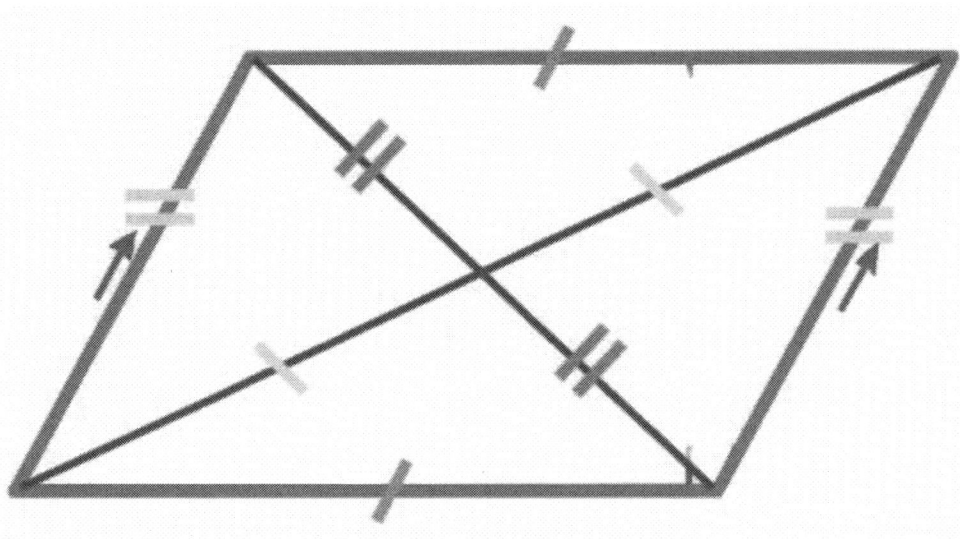

A Parallelogram

These properties provide multiple ways to prove that a quadrilateral is a parallelogram. A quadrilateral is a parallelogram if any of the following conditions is satisfied:
1. Both pairs of opposite sides are parallel.
2. Both pairs of opposite sides are congruent.
3. Both pairs of opposite angles are congruent.
4. The diagonals bisect each other.
5. One pair of opposite sides is both parallel and congruent.

In coordinate geometry, these conditions can be verified using the distance formula, midpoint formula, and slope formula. For instance, to verify that opposite sides are congruent, calculate the distances between the appropriate

51

vertices. To verify that diagonals bisect each other, calculate the midpoints of the diagonals and confirm they are the same point.

The area of a parallelogram can be calculated using the formula:

$$A = bh$$

where b is the length of the base and h is the height (the perpendicular distance from the base to the opposite side).

Alternatively, the area can be calculated using the formula:

$$A = |AD \times AB \sin C|$$

where |AD| and |AB| are the lengths of two adjacent sides, and C is the angle between them.

In a coordinate system, the area of a parallelogram can be calculated using the cross product of two adjacent sides or using the Shoelace formula if all vertices are known.

Rectangles, Rhombuses, and Squares

A rectangle is a parallelogram with four right angles. As a parallelogram, it inherits all the properties of parallelograms: opposite sides are parallel and congruent, opposite angles are congruent, consecutive angles are supplementary, and diagonals bisect each other. Additionally, rectangles have the following special properties:

1. All four angles are right angles (90°).
2. The diagonals are congruent:

$$AC = BD$$

3. The area of a rectangle can be calculated using:

$$A = l \times w$$

where l is the length and w is the width.

A rhombus is a parallelogram with four congruent sides. It inherits all the properties of parallelograms and has the following additional properties:

1. All four sides are congruent:

$$AB = BC = CD = DA$$

52

2. The diagonals bisect the interior angles:

$$\angle AOB = \angle BOC = \angle COD = \angle DOA$$

where O is the point of intersection of the diagonals.

3. The diagonals are perpendicular to each other:

$$AC \perp BD$$

4. The area of a rhombus can be calculated using:

$$A = \frac{1}{2} \times d_1 \times d_2$$

where d_1 and d_2 are the lengths of the diagonals.

A square is both a rectangle and a rhombus, combining the properties of both. It is a parallelogram with four right angles and four congruent sides. Squares have the following complete set of properties:

1. All four sides are congruent:

$$AB = BC = CD = DA$$

2. All four angles are right angles (90°).
3. Opposite sides are parallel:

$$AB \parallel DC \text{ and } AD \parallel BC$$

4. The diagonals are congruent:

$$AC = BD$$

5. The diagonals bisect each other:

$$AO = OC \text{ and } BO = OD$$

6. The diagonals are perpendicular to each other:

$$AC \perp BD$$

7. The diagonals bisect the interior angles.

The area of a square can be calculated using:

$$A = s^2$$

where s is the length of a side, or using:

$$A = \frac{1}{2} \times d^2$$

Where d is the length of a diagonal.

These special quadrilaterals form a hierarchy: all squares are rhombuses and rectangles, all rhombuses and rectangles are parallelograms, but not all parallelograms are rhombuses or rectangles, and not all rhombuses or rectangles are squares.

In coordinate geometry, these quadrilaterals can be identified by placing them in a coordinate system and verifying their properties using coordinate methods. For example, a rectangle aligned with the coordinate axes might have vertices at (0, 0), (a, 0), (a, b), and (0, b), where a and b are positive real numbers.

Trapezoids and Kites

Trapezoids and kites are quadrilaterals with specific properties that distinguish them from parallelograms and other quadrilaterals.

A trapezoid is a quadrilateral with exactly one pair of parallel sides, called the bases. The non-parallel sides are called the legs. If the legs are congruent, the trapezoid is isosceles. The parallel sides are typically denoted as a and c, and the height h is the perpendicular distance between them.

The area of a trapezoid can be calculated using:

$$A = \frac{1}{2} \times h \times (a + c)$$

where a and c are the lengths of the parallel sides (bases), and h is the height.

A midsegment of a trapezoid is a segment that connects the midpoints of the non-parallel sides. The midsegment is parallel to the bases, and its length equals the average of the lengths of the bases:

$$m = \frac{a + c}{2}$$

In an isosceles trapezoid, where the non-parallel sides are congruent, several additional properties emerge:

1. The angles along the same base are congruent:

$$m\angle A = m\angle B \text{ and } m\angle C = m\angle D$$

2. The diagonals are congruent:

$$AC = BD$$

3. A cyclic trapezoid, which can be inscribed in a circle, has the property that opposite angles are supplementary:

$$m\angle A + m\angle C = m\angle B + m\angle D = 180°$$

A kite is a quadrilateral with two pairs of adjacent congruent sides. If the adjacent congruent sides are $AB = AD$ and $BC = DC$, then several properties follow:

1. One diagonal (BD in this case) bisects the other diagonal (AC):

$$AO = OC$$

where O is the point of intersection of the diagonals.

2. One diagonal (AC in this case) bisects the angles at its endpoints:

$$\angle BAC = \angle DAC \text{ and } \angle BCA = \angle DCA$$

3. The diagonals are perpendicular to each other:

$$AC \perp BD$$

4. The area of a kite can be calculated using:

$$A = \frac{1}{2} \times d_1 \times d_2$$

where d_1 and d_2 are the lengths of the diagonals.

In coordinate geometry, trapezoids and kites can be analyzed by placing their vertices in a coordinate system. For example, a trapezoid might have vertices at $(0,0)$, $(a,0)$, (b,c), and (d,c), where the y-coordinates of the second pair of vertices are equal, indicating parallel sides.

It's important to note that while a rhombus is a special case of a kite (where all four sides are congruent), a trapezoid is distinct from a parallelogram because it has exactly one pair of parallel sides, whereas a parallelogram has two pairs.

Regular Polygons

A regular polygon is a polygon with all sides congruent and all interior angles congruent. Regular polygons possess a high degree of symmetry and special properties that make them important in geometry and various applications.

Every regular n-sided polygon has the following properties:

1. All n sides are congruent.
2. All n interior angles are congruent, each measuring:

$$m\angle = \frac{(n-2) \times 180°}{n}$$

3. All n exterior angles are congruent, each measuring:

$$m\angle_{ext} = \frac{360°}{n}$$

4. The sum of all interior angles equals $(n-2) \times 180°$.
5. The sum of all exterior angles equals $360°$.
6. Regular polygons have both rotational and reflectional symmetry. A regular n-sided polygon has:
- n axes of reflectional symmetry
- Rotational symmetry of order n (it can be rotated by multiples of 360°/n to coincide with itself)

The center of a regular polygon is the point equidistant from all vertices. This point is also the center of the circumscribed circle (the circle passing through all vertices) and the inscribed circle (the circle tangent to all sides).

Several measurements are associated with regular polygons:

The radius R of the circumscribed circle is the distance from the center to any vertex.

The apothem a is the distance from the center to the midpoint of any side. It represents the radius of the inscribed circle and can be calculated using:

$$a = R\cos\left(\frac{180°}{n}\right)$$

The side length s of a regular n-sided polygon with circumradius R can be calculated using:

$$s = 2R sin\left(\frac{180°}{n}\right)$$

The perimeter P of a regular n-sided polygon with side length s is:

$$P = n \times s$$

The area A of a regular n-sided polygon can be calculated using several equivalent formulas:

$$A = \frac{1}{2} \times n \times s \times a$$

$$A = \frac{1}{2} \times P \times a$$

$$A = \frac{1}{4} \times n \times s^2 \times \left(\frac{180°}{n}\right)$$

$$A = \frac{1}{2} \times n \times R^2 \times \left(\frac{360°}{n}\right)$$

As the number of sides of a regular polygon increases, the polygon approaches a circle. The perimeter approaches the circumference $2\pi R$, and the area approaches πR^2.

Regular polygons can be constructed using a compass and straightedge for certain numbers of sides (3, 4, 5, 6, 8, 10, 12, 15, 16, etc.). It has been proven that a regular n-sided polygon is constructible with compass and straightedge if and only if n is a product of a power of 2 and distinct Fermat primes.

In coordinate geometry, regular polygons are often placed with their center at the origin. The vertices of a regular n-sided polygon with circumradius R can be calculated using:

$$\left(R cos\left(\frac{2\pi k}{n}\right), R\backslash sin\left(\frac{2\pi k}{n}\right)\right)$$

for $k = 0, 1, 2, \ldots, n - 1$.

Practice Problems

1. The sum of the interior angles of a convex polygon is 1440°. How many sides does the polygon have?

2. The vertices of quadrilateral PQRS are P(0, 0), Q(4, 0), R(4, 4), and S(0, 4). Prove that PQRS is a square by verifying that all four sides are congruent and all four angles are right angles.

Solutions

1. To find the number of sides of a polygon given the sum of interior angles:

Given: The sum of interior angles is 1440°.

Recall that the sum of interior angles in an n-sided polygon is: $S = (n-2) \times 180°$

Substituting the given information: $1440° = (n-2) \times 180°$

Dividing both sides by 180°: $\frac{1440°}{180°} = n - 2, 8 = n - 2, n = 10$

The polygon has 10 sides.

2. To prove that quadrilateral PQRS with vertices P(0, 0), Q(4, 0), R(4, 4), and S(0, 4) is a square:

First, calculate the lengths of all four sides:

$$PQ = \sqrt{[(4-0)^2 + (0-0)^2]} = 4 \quad QR = \sqrt{[(4-4)^2 + (4-0)^2]} = 4 \quad RS = \sqrt{[(0-4)^2 + (4-4)^2]}$$
$$= 4 \quad SP = \sqrt{[(0-0)^2 + (0-4)^2]} = 4$$

All four sides have the same length (4 units), so the quadrilateral has four congruent sides.

Next, verify that all angles are right angles using the slopes of adjacent sides:

Slope of $PQ = (0-0)/(4-0) = 0$ Slope of $QR = (4-0)/(4-4) = \infty$ (vertical line) Slope of $RS = (4-4)/(0-4) = 0$ $Slope\ of\ SP = (0-4)/(0-0) = \infty$ (vertical line)

Since adjacent sides have slopes that are either 0 or ∞, they are perpendicular to each other, forming right angles.

With four congruent sides and four right angles, PQRS is a square.

RIGHT TRIANGLES AND TRIGONOMETRY

The Pythagorean Theorem

The Pythagorean Theorem is one of the most fundamental relationships in geometry, establishing a connection between the sides of a right triangle. The theorem states that in a right triangle, the square of the length of the hypotenuse equals the sum of the squares of the lengths of the other two sides, commonly known as the legs.

If a and b represent the lengths of the legs, and c represents the length of the hypotenuse, then:

$$a^2 + b^2 = c^2$$

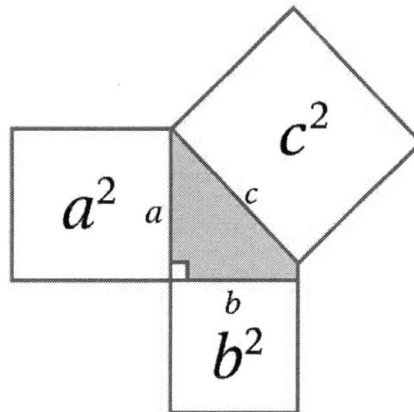

A Right-angled triangle

This theorem is named after the ancient Greek mathematician Pythagoras, although evidence suggests it was known to earlier civilizations. The theorem can be proven in numerous ways, including algebraic methods, similarity principles, and area relationships.

One elegant proof involves showing that the sum of the areas of squares constructed on the legs equals the area of the square constructed on the hypotenuse. Another approach uses similar triangles formed by the altitude to the hypotenuse.

The Pythagorean Theorem applies exclusively to right triangles, serving as both a defining characteristic and a powerful analytical tool. It enables the calculation of unknown side lengths when two sides are known:

$$c = \sqrt{a^2 + b^2}$$

$$a = \sqrt{c^2 - b^2}$$

$$b = \sqrt{c^2 - a^2}$$

The converse of the Pythagorean Theorem is also true: if the sides of a triangle satisfy the equation $a^2 + b^2 = c^2$, then the triangle is a right triangle with the right angle opposite to side c.

A Pythagorean triple consists of three positive integers a, b, and c that satisfy the Pythagorean equation. The most familiar example is (3, 4, 5), since $3^2 + 4^2 = 9 + 16 = 25 = 5^2$. Other examples include (5, 12, 13), (8, 15, 17), and (7, 24, 25). Any multiple of a Pythagorean triple is also a Pythagorean triple; for example, (6, 8, 10) is a multiple of (3, 4, 5).

The general formula for generating Pythagorean triples is:

$$a = m^2 - n^2, b = 2mn, c = m^2 + n^2$$

where m and n are positive integers with m > n, and m and n are coprime (their greatest common divisor is 1) and not both odd.

The Pythagorean Theorem extends to higher dimensions through the distance formula. In three-dimensional space, the distance between points (x_1, y_1, z_1) and (x_2, y_2, z_2) is:

$$d = \sqrt{(x_2 - x_1)^2 + (y_2 - y_1)^2 + (z_2 - z_1)^2}$$

Special Right Triangles

Certain right triangles have special angle measures that yield consistent relationships between their sides. These special right triangles—the 30°-60°-90° triangle and the 45°-45°-90° triangle—appear frequently in geometry and trigonometry problems, making their properties worth memorizing.

The 45°-45°-90° triangle, also called the isosceles right triangle, has two angles of 45° and one right angle (90°). The two legs of this triangle are congruent because they are opposite to equal angles. If the length of each leg is denoted as s, then the length of the hypotenuse is $s\sqrt{2}$.

This relationship follows directly from the Pythagorean Theorem:

$$s^2 + s^2 = c^2, 2s^2 = c^2, c = s\sqrt{2}$$

The 45°-45°-90° triangle exhibits a ratio of side lengths:

$$leg : leg : hypotenuse = 1 : 1 : \sqrt{2}$$

For example, in a 45°-45°-90° triangle with legs of length 5 units, the hypotenuse would measure $5\sqrt{2} \approx 7.07$ units.

The 30°-60°-90° triangle has angles of 30°, 60°, and 90°. If the shortest leg (opposite to the 30° angle) has length s, then the hypotenuse has length 2s, and the remaining leg (opposite to the 60° angle) has length $s\sqrt{3}$.

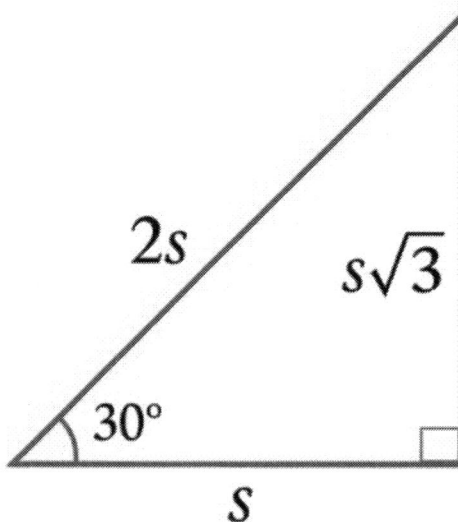

A 30-60-90 triangle

These relationships can be derived by considering an equilateral triangle with side length 2s. When an altitude is drawn from one vertex to the opposite side, it creates two congruent 30°-60°-90° triangles. The altitude has length $s\sqrt{3}$ (by the Pythagorean Theorem), and it divides the opposite side into two segments of length s each.

The 30°-60°-90° triangle exhibits a ratio of side lengths:

$$shortest\ leg : longer\ leg : hypotenuse = 1 : \sqrt{3} : 2$$

or example, in a 30°-60°-90° triangle with the shortest leg of length 3 units, the longer leg would measure $3\sqrt{3} \approx$ 5.2 units, and the hypotenuse would measure 6 units.

These special right triangles are particularly useful in trigonometry, as they provide exact values for trigonometric functions of common angles. They also appear frequently in geometric figures, such as regular polygons and three-dimensional solids.

61

In coordinate geometry, these triangles can be placed with vertices at specific points to create well-known configurations. For instance, a 45°-45°-90° triangle can be placed with vertices at $(0, 0)$, $(1, 0)$, and $(1, 1)$, while a 30°-60°-90° triangle can be placed with vertices at $(0, 0)$, $(1, 0)$, and $(1/2, \sqrt{3}/2)$.

Trigonometric Ratios

Trigonometric ratios establish relationships between the angles and sides of a right triangle, providing powerful tools for solving geometric problems. The three primary trigonometric ratios are sine, cosine, and tangent, defined for an acute angle θ in a right triangle as follows:

$$\sin \theta = \frac{\text{opposite}}{\text{hypotenuse}}$$

$$\cos \theta = \frac{\text{adjacent}}{\text{hypotenuse}}$$

$$\tan \theta = \frac{\text{opposite}}{\text{adjacent}} = \frac{\sin \theta}{\cos \theta}$$

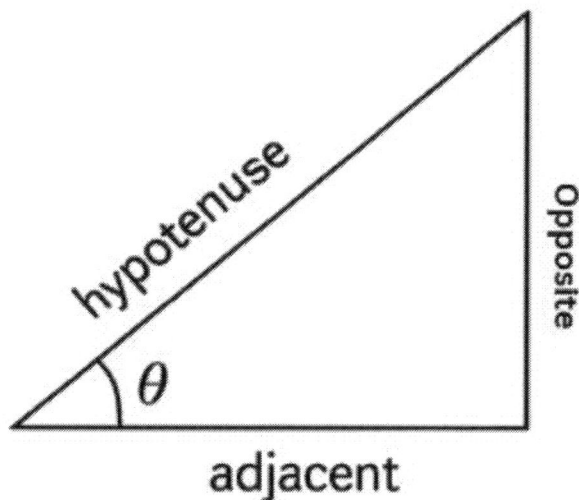

Right triangle

In these definitions, "opposite" refers to the side opposite to angle θ, "adjacent" refers to the side adjacent to angle θ (excluding the hypotenuse), and "hypotenuse" is the side opposite to the right angle.

These ratios depend solely on the measure of angle θ, not on the size of the triangle. This is because similar triangles have proportional sides, so the ratios of sides remain constant for a given angle.

The reciprocal trigonometric ratios—cosecant, secant, and cotangent—are defined as:

$$\csc\theta = \frac{1}{\sin\theta} = \frac{\text{hypotenuse}}{\text{opposite}}$$

$$\sec\theta = \frac{1}{\cos\theta} = \frac{\text{hypotenuse}}{\text{adjacent}}$$

$$\cot\theta = \frac{1}{\tan\theta} = \frac{\text{adjacent}}{\text{opposite}} = \frac{\cos\theta}{\sin\theta}$$

For the special angles discussed in the previous section, the trigonometric ratios have exact values:

For a 30° angle: $\sin 30° = \frac{1}{2}$, $\cos 30° = \frac{\sqrt{3}}{2}$, $\tan 30° = \frac{1}{\sqrt{3}} = \frac{\sqrt{3}}{3}$

For a 45° angle: $\sin 45° = \frac{1}{\sqrt{2}} = \frac{\sqrt{2}}{2}$, $\cos 45° = \frac{1}{\sqrt{2}} = \frac{\sqrt{2}}{2}$, $\tan 45° = 1$

For a 60° angle: $\sin 60° = \frac{\sqrt{3}}{2}$, $\cos 60° = \frac{1}{2}$, $\tan 60° = \sqrt{3}$

In a right triangle, the trigonometric ratios provide a means to find unknown side lengths and angle measures. If one side length and one acute angle are known, the remaining side lengths can be calculated using the appropriate trigonometric ratios. Similarly, if two side lengths are known, the acute angles can be determined using inverse trigonometric functions:

$$\theta = \sin^{-1}\left(\frac{\text{opposite}}{\text{hypotenuse}}\right)$$

$$\theta = \cos^{-1}\left(\frac{\text{adjacent}}{\text{hypotenuse}}\right)$$

$$\theta = \tan^{-1}\left(\frac{\text{opposite}}{\text{adjacent}}\right)$$

The Pythagorean identity, a fundamental relationship among trigonometric functions, states that for any angle θ:

$$\sin^2\theta + \cos^2\theta = 1$$

This identity can be derived from the Pythagorean Theorem applied to a right triangle with hypotenuse of length 1.

Applications of Trigonometry

One common application is finding heights and distances indirectly. When direct measurement is impractical, trigonometric ratios enable calculations based on accessible measurements and angles.

For example, to find the height of a building, one can measure the angle of elevation from a point on the ground to the top of the building, along with the distance from that point to the base of the building. If the angle of elevation is θ and the distance is d, then the height h of the building can be calculated using:

$$h = d \times \tan \theta$$

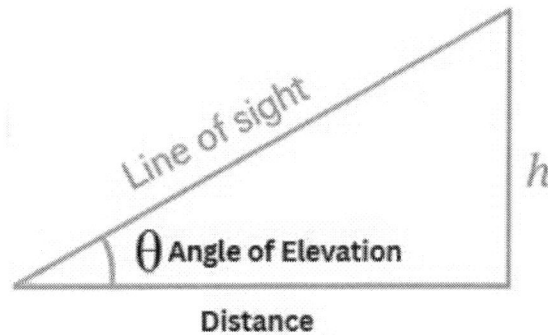

Angle of elevation with distance and height

Similarly, to find the distance to an inaccessible object, one can measure angles from two different observation points separated by a known distance. This technique, called triangulation, forms the basis of many surveying methods.

In navigation, trigonometry helps determine position and bearing. The course of a ship or aircraft can be calculated using angles and distances. If a ship travels distance d in direction θ from its starting point, its change in position along the east-west axis is $d \times \cos \theta$ and along the north-south axis is $d \times \sin \theta$.

Trigonometry also plays a role in analyzing periodic phenomena. The sine and cosine functions model cyclic behavior, such as waves, oscillations, and rotations. For instance, the displacement y of a point on a simple pendulum of length L as a function of time t can be approximated by:

$$y = L \times \sin\left(\sqrt{\frac{g}{L}} \times t\right)$$

where g is the acceleration due to gravity.

In coordinate geometry, trigonometric functions facilitate transformations like rotations. The coordinates of a point (x, y) after rotation by angle θ around the origin become:

$$(x', y') = (x \cos \theta - y \sin \theta, x \sin \theta + y \cos \theta)$$

Architectural design employs trigonometry to calculate structural elements like roof pitches, arches, and support systems. If a roof has a pitch angle of θ, and the width of the building is w, then the length of the roof rafter is:

$$r = \frac{w/2}{\cos \theta}$$

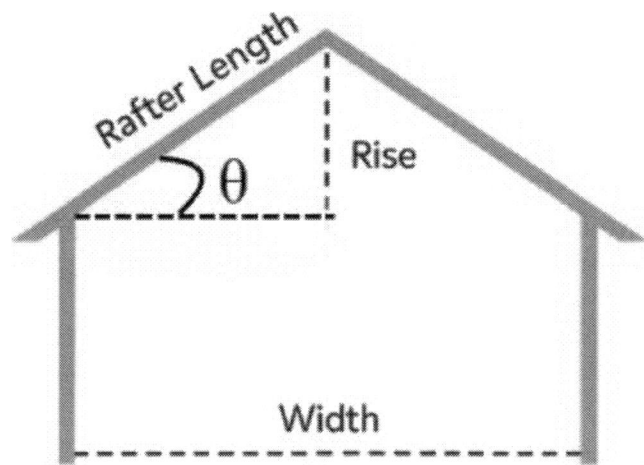

Roof diagram

Optics uses trigonometry to analyze the behavior of light, including refraction and reflection. Snell's law, which describes how light bends when passing from one medium to another, is expressed using trigonometric functions:

$$\frac{\sin \theta_1}{\sin \theta_2} = \frac{v_1}{v_2} = \frac{n_2}{n_1}$$

where θ_1 and θ_2 are the angles of incidence and refraction, v_1 and v_2 are the velocities of light in the respective media, and n_1 and n_2 are the refractive indices.

Physics applications include decomposing vectors into components, analyzing forces on inclined planes, and describing wave phenomena. For a force F acting at angle θ to the horizontal, the horizontal component is $F \cos \theta$ and the vertical component is $F \sin \theta$.

Law of Sines and Cosines

While the trigonometric ratios discussed earlier apply specifically to right triangles, the Laws of Sines and Cosines extend trigonometric principles to any triangle, providing powerful tools for solving problems involving oblique triangles (triangles without a right angle).

The Law of Sines states that the ratio of the length of a side to the sine of the opposite angle is constant for all three sides of a triangle. For a triangle with sides a, b, and c, and opposite angles A, B, and C, respectively:

$$\frac{a}{\sin A} = \frac{b}{\sin B} = \frac{c}{\sin C} = 2R$$

where R is the radius of the circumscribed circle of the triangle.

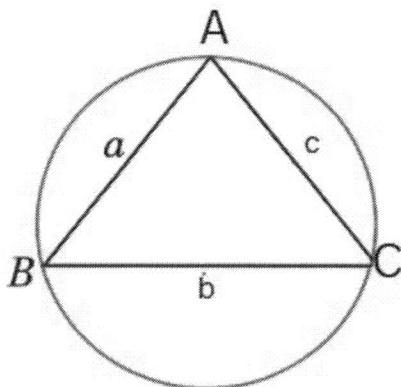

Triangle circumscribed in a circle

This law can be derived using the area formula for a triangle and the properties of the circumscribed circle. It provides a method for finding unknown sides or angles when certain combinations of sides and angles are known:

1. When two angles and one side are known (AAS or ASA case)
2. When two sides and the angle opposite to one of them are known (SSA case, which may yield two solutions or no solution)

For example, if angles A and B and side a are known, side b can be calculated using:

$$b = a \times \frac{\sin B}{\sin A}$$

The Law of Cosines relates the square of the length of a side to the squares of the lengths of the other two sides and the cosine of the included angle. For a triangle with sides a, b, and c, and angles A, B, and C:

$$a^2 = b^2 + c^2 - 2bc \cos A$$

$$b^2 = a^2 + c^2 - 2ac \cos B$$

$$c^2 = a^2 + b^2 - 2ab \cos C$$

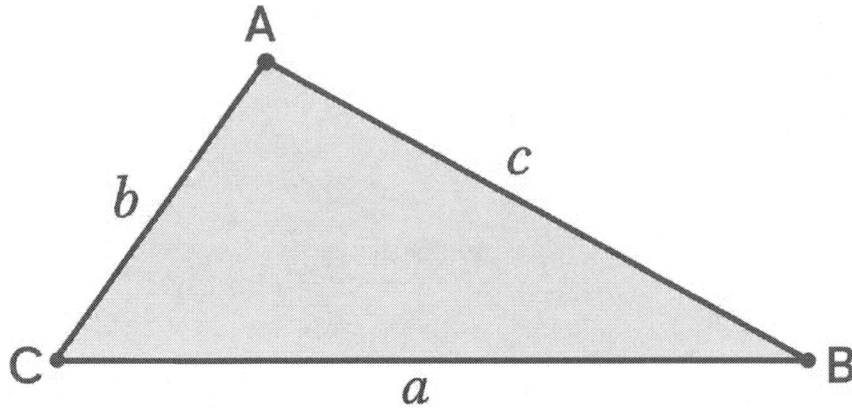

LAW OF COSINES

$$c^2 = a^2 + b^2 - 2 \cdot a \cdot b \cdot \cos(C)$$

The Law of Cosines generalizes the Pythagorean Theorem. If angle C is 90°, then cos C = 0, and the third equation reduces to $c^2 = a^2 + b^2$, which is the Pythagorean Theorem.

This law is particularly useful for:

1. Finding the third side of a triangle when two sides and the included angle are known (SAS case)
2. Finding an angle when all three sides are known (SSS case)

For example, if sides b, c, and angle A are known, side a can be calculated using:

$$a = \sqrt{b^2 + c^2 - 2bc \cos A}$$

If all three sides are known, angle A can be calculated using:

$$A = \cos^{-1}\left(\frac{b^2 + c^2 - a^2}{2bc}\right)$$

Together, the Laws of Sines and Cosines provide complete methods for solving triangles. The appropriate law to use depends on the given information:

- Law of Sines: Use when given AAS, ASA, or SSA
- Law of Cosines: Use when given SAS or SSS

In the SSA case, the Law of Sines may yield zero, one, or two solutions, depending on whether the given values can form a valid triangle. This ambiguity is known as the ambiguous case of the Law of Sines.

Practice Problems

1. In a 30°-60°-90° triangle, the shortest side measures 8 inches. Find the lengths of the other two sides.
2. From a point 150 feet from the base of a building, the angle of elevation to the top of the building is 32°. Find the height of the building to the nearest foot.
3. A ladder 10 feet long leans against a vertical wall, making an angle of 65° with the ground. How high up the wall does the ladder reach? Round to the nearest tenth of a foot.
4. In triangle ABC, a = 8 cm, b = 12 cm, and c = 15 cm. Find the measure of angle A to the nearest degree.

Solutions

1. In a 30°-60°-90° triangle with the shortest side measuring 8 inches:

The shortest side is opposite to the 30° angle. Using the properties of 30°-60°-90° triangles:

The hypotenuse (opposite to the 90° angle) has length: hypotenuse $= 2 \times$ shortest side $= 2 \times 8 = 16$ inches

The remaining side (opposite to the 60° angle) has length: longer leg $=$ shortest side $\times \sqrt{3} = 8 \times \sqrt{3} \approx$ 13.86 inches

The two sides measure 16 inches and $8\sqrt{3}$ inches (approximately 13.86 inches).

2. To find the height of a building when the angle of elevation is 32° from a point 150 feet from the base:

Using the tangent function: $\tan 3\,2° = \frac{\text{height}}{\text{distance}}$ height $=$ distance $\times \tan 3\,2°$height $= 150$ feet $\times \tan 3\,2°$height $= 150$ feet $\times$ height $= 93.735$ feet

Rounding to the nearest foot, the height of the building is 94 feet.

3. To find how high up a wall a 10-foot ladder reaches when it makes an angle of 65° with the ground:

Using the sine function: $\sin 6\,5° = \frac{\text{height}}{\text{ladder length}}$ height $=$ ladder length $\times \sin 6\,5°$ height $= 10$ feet $\times \sin 6\,5°$ height $= 10$ feet $\times 0.9063$ height $= 9.063$ feet

Rounding to the nearest tenth, the ladder reaches 9.1 feet up the wall.

4. To find angle A in triangle ABC with a = 8 cm, b = 12 cm, and c = 15 cm:

Using the Law of Cosines: $a^2 = b^2 + c^2 - 2bc \cos$ $\cos A = \frac{b^2+c^2-a^2}{2bc}$ $\cos A = \frac{12^2+15^2-8^2}{2\times12\times15}$ $\cos A =$ $\frac{144+225-64}{360}$ $\cos A = \frac{305}{360} \approx$ $A = \cos^{-1}(0.8472) \approx 32.$

Rounding to the nearest degree, angle A measures 32°.

CIRCLES

Properties of Circles

A circle is the set of all points in a plane that are equidistant from a fixed point called the center. The distance from the center to any point on the circle is called the radius. A circle with center at point O and radius r can be denoted as circle O or C(O, r).

Several fundamental terms and elements are associated with circles:

1. A diameter is a line segment that passes through the center of the circle and has its endpoints on the circle. The length of a diameter is twice the radius: $d = 2r$.
2. A chord is a line segment with endpoints on the circle. A diameter is the longest possible chord.
3. A secant is a line that intersects a circle at exactly two points.
4. A tangent is a line that intersects a circle at exactly one point, called the point of tangency. A tangent is perpendicular to the radius drawn to the point of tangency.

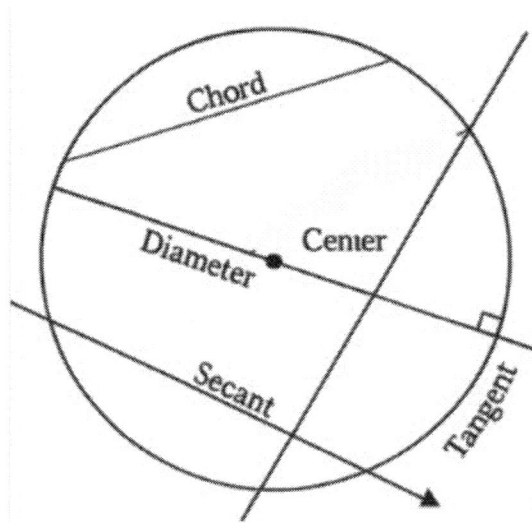

An arc is a portion of the circumference of a circle. A minor arc is less than a semicircle and is denoted by the endpoints, such as $\overgroup{AB}$. A major arc is greater than a semicircle and is denoted by the endpoints and an intermediate point, such as $\overgroup{ACB}$.

A central angle is an angle with its vertex at the center of the circle and its sides containing radii of the circle. The measure of an arc equals the measure of its central angle.

The circumference of a circle—the distance around the circle—is given by:

$$C = 2\pi r = \pi d$$

where r is the radius and d is the diameter.

The area of a circle is given by:

$$A = \pi r^2 = \frac{\pi d^2}{4}$$

A sector of a circle is the region bounded by two radii and the arc between them. If the central angle of the sector is θ (in radians), the area of the sector is:

$$A_{sector} = \frac{1}{2} r^2$$

If the central angle is measured in degrees, the formula becomes:

$$A_{sector} = \frac{\theta}{360°} \times \pi r^2$$

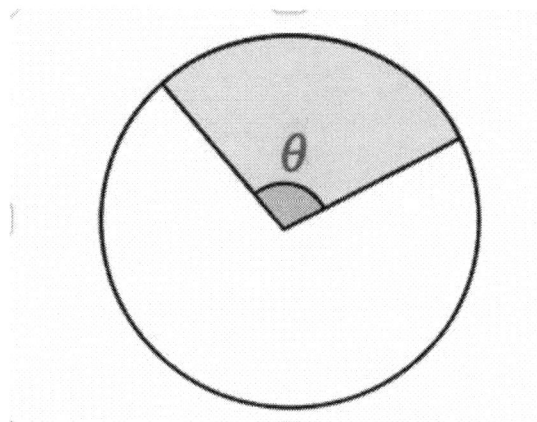

Circle showing a sector with central angle θ

A segment of a circle is the region bounded by a chord and the arc it cuts off. The area of a segment can be calculated by subtracting the area of a triangle from the area of a corresponding sector.

The power of a point P with respect to a circle is a measure of how far the point is from the circle. If P is outside the circle and a tangent from P touches the circle at point T, then the power of P is the square of the length of the tangent segment PT. This value is the same for any tangent drawn from P to the circle.

The power of a point theorem states that if a point P is outside a circle and lines through P intersect the circle at points A and B, and C and D, respectively, then:

$$PA \times PB = PC \times PD$$

This relationship holds true regardless of which secant lines through P are chosen.

Circles can be placed in a coordinate system, where they can be represented by an equation. The standard form of the equation of a circle with center (h, k) and radius r is:

$$(x - h)^2 + (y - k)^2 = r^2$$

The general form of the equation of a circle is:

$$x^2 + y^2 + Dx + Ey + F = 0$$

where the center is at $(-D/2, -E/2)$ and the radius is $\sqrt{(D/2)^2 + (E/2)^2 - F}$.

Arc Length and Sector Area

Arc length and sector area calculations are extensions of the circumference and area formulas for circles, adjusted for portions of a circle rather than the entire circle.

The length of an arc depends on the radius of the circle and the measure of the central angle that subtends the arc. If θ is the measure of the central angle in radians, and r is the radius of the circle, then the arc length s is given by:

$$s = r\theta$$

If the central angle is measured in degrees, the formula becomes:

$$s = \frac{2\pi r \cdot \theta}{360°}$$

This formula can be derived by considering the ratio of the arc length to the circumference of the entire circle, which equals the ratio of the central angle to 360°:

$$\frac{s}{2\pi r} = \frac{\theta}{360°}$$

For example, if a circle has a radius of 5 units and an arc subtends a central angle of 60°, the arc length is:

$$s = \frac{2\pi \cdot 5 \cdot 60°}{360°} = \frac{10\pi \cdot 60°}{360°} = \frac{10\pi}{6} = \frac{5\pi}{3} \approx 5.24 \text{ units}$$

A sector of a circle is the region bounded by two radii and the arc between them. The area of a sector is proportional to the central angle. If θ is the measure of the central angle in radians, and r is the radius of the circle, then the area of the sector is:

71

$$A_{sector} = \frac{1}{2}r^2$$

If the central angle is measured in degrees, the formula becomes:

$$A_{sector} = \frac{\pi r^2 \cdot \theta}{360°}$$

This formula can be derived by considering the ratio of the sector area to the area of the entire circle, which equals the ratio of the central angle to 360°:

$$\frac{A_{sector}}{\pi r^2} = \frac{\theta}{360°}$$

For example, if a circle has a radius of 4 units and a sector has a central angle of 45°, the area of the sector is:

$$A_{sector} = \frac{\pi \cdot 4^2 \cdot 45°}{360°} = \frac{16\pi \cdot 45°}{360°} = \frac{16\pi}{8} = 2\pi \approx 6.28 \text{ square units}$$

A related concept is the area of a segment, which is the region bounded by a chord and the arc it subtends. The area of a segment can be calculated by subtracting the area of the triangle formed by the center of the circle and the chord from the area of the corresponding sector:

$$A_{segment} = A_{sector} - A_{triangle}$$

If the central angle is θ and the radius is r, then:

$$A_{segment} = \frac{r^2}{2}(\theta - \sin\theta)$$

where θ is measured in radians.

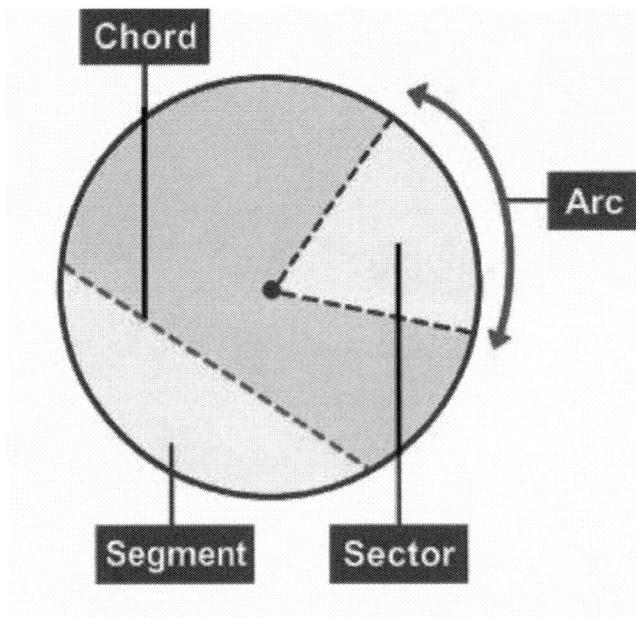

Circle showing segment area with chord and arc

Angle Relationships in Circles

A central angle has its vertex at the center of the circle and its sides contain radii of the circle. The measure of an arc equals the measure of its central angle:

$$m\widehat{AB} = m\angle AOB$$

where O is the center of the circle.

An inscribed angle has its vertex on the circle and its sides contain chords of the circle. The inscribed angle theorem states that the measure of an inscribed angle equals half the measure of its intercepted arc:

$$m\angle ACB = \frac{1}{2}m\widehat{AB}$$

A corollary to this theorem is that inscribed angles that intercept the same arc are congruent. Also, an angle inscribed in a semicircle is a right angle.

An angle formed by a tangent and a chord has its vertex at the point of tangency. The measure of such an angle equals half the measure of the intercepted arc:

$$m\angle PQR = \frac{1}{2}m\widehat{PR}$$

73

where PQ is the tangent and QR is the chord.

Two chords that intersect inside a circle form two pairs of vertical angles. The measure of each angle equals half the sum of the measures of the arcs intercepted by the angle and its vertical angle:

$$m\angle APC = \frac{1}{2}\left(m\widehat{AB} + m\widehat{CD}\right)$$

where P is the point of intersection of chords AC and BD.

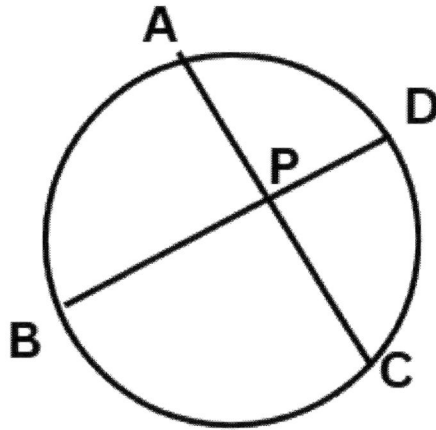

Circle showing intersecting chords AC and BD

An angle formed by two secants, two tangents, or a secant and a tangent from an external point has a special relationship with the intercepted arcs. If the angle is formed outside the circle, its measure equals half the difference of the measures of the intercepted arcs:

$$m\angle APB = \frac{1}{2}\left|m\widehat{AB} - m\widehat{CD}\right|$$

where P is the external point, and the lines through P intersect the circle at points A, B, C, and D.

Tangent Properties

Tangent lines to circles have several important properties that can be applied to solve geometric problems and The fundamental property of a tangent line is that it is perpendicular to the radius drawn to the point of tangency. If line ℓ is tangent to a circle at point P, and O is the center of the circle, then:

$$\ell \perp \overline{OP}$$

This property can be proven by considering the distances from points on the line to the center of the circle. The distance from a point to the center is minimized at the point of tangency.

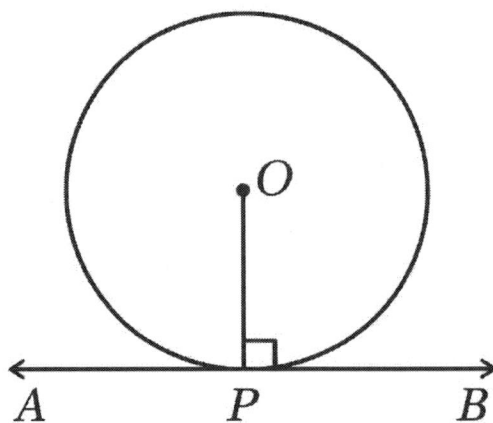

Circle showing tangent perpendicular to radius at point of tangency

From this fundamental property, several other important properties can be derived:

If two tangent lines are drawn to a circle from an external point P, then the two tangent segments (from P to the points of tangency) are congruent:

$$\overline{PT}_1 \cong \overline{PT}_2$$

where T_1 and T_2 are the points of tangency.

The line segment from the external point P to the center O of the circle bisects the angle formed by the two tangent segments:

$$\angle T_1 P T_2 \text{ is bisected by } \overline{PO}$$

If two circles are tangent (they touch at exactly one point), then the line joining their centers passes through their point of tangency. This is true for both external tangency (when the circles lie outside each other) and internal tangency (when one circle lies inside the other).

The common external tangents to two non-concentric circles are congruent. Similarly, the common internal tangents (if they exist) are congruent.

A tangent-secant theorem states that if a tangent and a secant are drawn to a circle from an external point P, then the square of the length of the tangent segment equals the product of the secant segment and its external part:

$$\overline{PT}^2 = \overline{PA} \cdot \overline{PB}$$

where T is the point of tangency, and A and B are the points where the secant intersects the circle.

This is a special case of the power of a point theorem, which states that for any secants drawn from an external point P to a circle, the products of the secant segments and their external parts are equal:

$$\overline{PA_1} \cdot \overline{PB_1} = \overline{PA_2} \cdot \overline{PB_2}$$

These properties of tangents are used in various geometric constructions and proofs. For example, tangent properties can be used to construct tangent lines from an external point to a circle, to find the center of a circle, or to solve problems involving tangent circles.

In coordinate geometry, the condition for a line to be tangent to a circle can be expressed algebraically. If a line with equation $ax + by + c = 0$ is tangent to a circle with center (h, k) and radius r, then the distance from the center to the line equals the radius:

$$\frac{|ah + bk + c|}{\sqrt{a^2 + b^2}} = r$$

Equations of Circles

The equation of a circle in a coordinate system provides an algebraic representation of the circle, enabling the application of algebraic methods to geometric problems involving circles.

The standard form of the equation of a circle with center (h, k) and radius r is:

$$(x - h)^2 + (y - k)^2 = r^2$$

This equation expresses the fact that every point (x, y) on the circle is at a distance r from the center (h, k), according to the distance formula.

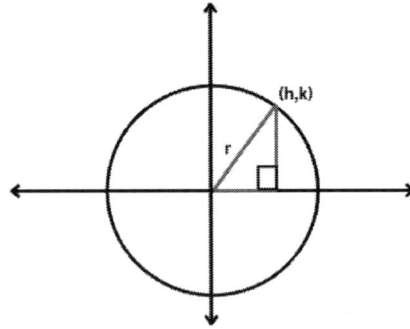

Coordinate grid showing circle with center (h,k) and radius r

For example, the equation of a circle with center (3, -2) and radius 5 is:

$$(x - 3)^2 + \left(y - (-2)\right)^2 = 5^2$$

$$(x - 3)^2 + (y + 2)^2 = 25$$

Special cases include:

- A circle with center at the origin (0, 0) has the equation: $x^2 + y^2 = r^2$

- A circle with center at (h, 0) has the equation: $(x - h)^2 + y^2 = r^2$

- A circle with center at (0, k) has the equation: $x^2 + (y - k)^2 = r^2$

The standard form can be expanded to obtain the general form of the equation of a circle:

$$x^2 + y^2 + Dx + Ey + F = 0$$

where D, E, and F are constants.

To convert from general form to standard form (a process called completing the square), group the terms with x and y separately:

$$(x^2 + Dx) + (y^2 + Ey) + F = 0$$

Complete the square for both the x and y terms:

$$\left(x^2 + Dx + \frac{D^2}{4}\right) + \left(y^2 + Ey + \frac{E^2}{4}\right) + F - \frac{D^2}{4} - \frac{E^2}{4} = 0$$

which simplifies to:

$$\left(x + \frac{D}{2}\right)^2 + \left(y + \frac{E}{2}\right)^2 = \frac{D^2}{4} + \frac{E^2}{4} - F$$

This gives the center at $\left(-\frac{D}{2}, -\frac{E}{2}\right)$ and radius $\sqrt{\frac{D^2}{4} + \frac{E^2}{4} - F}$.

For example, to convert the equation $x^2 + y^2 + 6x - 4y + 12 = 0$ to standard form:

$$(x^2 + 6x) + (y^2 - 4y) + 12 = (x^2 + 6x + 9) + (y^2 - 4y + 4) + 12 - 9 - 4 = (x + 3)^2 + (y - 2)^2 = 1$$

This is the equation of a circle with center (-3, 2) and radius 1.

Not every equation of the form $x^2 + y^2 + Dx + Ey + F = 0$ represents a circle. If $\frac{D^2}{4} + \frac{E^2}{4} - F < 0$, there are no points satisfying the equation, so it represents an imaginary circle. If $\frac{D^2}{4} + \frac{E^2}{4} - F = 0$, the equation represents a single point, sometimes called a circle of zero radius.

In the coordinate plane, several relationships involving circles can be expressed and analyzed using their equations:

- Two circles are concentric if they have the same center.
- Two circles intersect if the distance between their centers is less than the sum of their radii but greater than the absolute difference of their radii.
- Two circles are tangent externally if the distance between their centers equals the sum of their radii.
- Two circles are tangent internally if the distance between their centers equals the absolute difference of their radii.
- A line is tangent to a circle if the distance from the center of the circle to the line equals the radius.

The equation of a circle can be used to find points of intersection with lines or other circles, to determine whether a point lies on, inside, or outside the circle, and to analyze various geometric relationships involving the circle.

Practice Problems

1. Find the equation of a circle with center (-2, 3) and passing through the point (1, -1).
2. Determine the center and radius of the circle with equation $x^2 + y^2 - 4x + 6y + 4 = 0$.
3. In circle O, chord AB is 8 units long and is 3 units from the center O. Find the radius of the circle.
4. Two tangent segments are drawn to a circle from an external point P. If each tangent segment has length 12 units and the center of the circle is 15 units from P, find the radius of the circle.

5. A sector of a circle has a central angle of 45° and an area of 9π square units. Find the radius of the circle and the length of the arc that bounds the sector.

Solutions

1. To find the equation of a circle with center (-2, 3) passing through (1, -1):

The standard form of the equation is: $(x - h)^2 + (y - k)^2 = r^2$

Where (h, k) is the center and r is the radius.

The center is (-2, 3), so h = -2 and k = 3.

To find the radius, calculate the distance from the center to the point (1, -1): $r = \sqrt{\left(1 - (-2)\right)^2 + (-1 - 3)^2} =$

$\sqrt{3^2 + (-4)^2} = \sqrt{9 + 16} = \sqrt{25} = 5$

The equation of the circle is: $\left(x - (-2)\right)^2 + (y - 3)^2 = 5^2 \ (x + 2)^2 + (y - 3)^2 = 25$

2. To determine the center and radius of the circle with equation $x^2 + y^2 - 4x + 6y + 4 = 0$:

Rearrange to group terms with x and y: $(x^2 - 4x) + (y^2 + 6y) + 4 = 0$

Complete the square for both x and y terms: $(x^2 - 4x + 4) + (y^2 + 6y + 9) + 4 - 4 - 9 = 0 \ (x - 2)^2 +$ $(y + 3)^2 = 9$

The circle has center (2, -3) and radius 3.

3. To find the radius of circle O where chord AB is 8 units long and 3 units from the center:

Let the radius of the circle be r, and let the distance from the center O to the chord AB be d = 3 units.

Draw a perpendicular from O to the chord AB, meeting AB at point M. This creates a right triangle OMA where OM = d = 3 and MA = AB/2 = 4 units.

Using the Pythagorean Theorem in triangle OMA: $r^2 = d^2 + (AB/2)^2 \ r^2 = 3^2 + 4^2 \ r^2 = 9 + 16 \ r^2 = 25 \ r = 5$

The radius of the circle is 5 units.

4. To find the radius of a circle given that tangent segments from point P are 12 units long and the center is 15 units from P:

Let r be the radius of the circle and let d be the distance from P to the center O (d = 15).

If PT is a tangent segment from P to the circle at point T, then PT = 12 and $\angle OTP = 90°$ (tangent perpendicular to radius).

In right triangle OTP: $d^2 = r^2 + PT^2 \ 15^2 = r^2 + 12^2 \ 225 = r^2 + 144 \ r^2 = 225 - 144 \ r^2 = r =$

The radius of the circle is 9 units.

5. To find the radius and arc length for a sector with central angle 45° and area 9π square units:

For a sector with central angle θ (in degrees) and radius r: $A_{sector} = \dfrac{\theta}{360°} \times \pi r^2$

Given that A sector = 9π and θ = 45°: $9\pi = \dfrac{45°}{360°} \times \pi r^2 \ 9\pi = \dfrac{1}{8} \times \pi r^2 \ 9 = \dfrac{r^2}{8} \ r^2 = 72 \ r = 6\sqrt{2} \approx 8.49$

The arc length s is: $s = \frac{\theta}{360°} \times 2\pi r = \frac{45°}{360°} \times 2\pi \times 6\sqrt{2} = \frac{12\pi\sqrt{2}}{8} = \frac{3\pi\sqrt{2}}{2} \approx 6.67$

The radius is $6\sqrt{2}$ units and the arc length is $\frac{3\pi\sqrt{2}}{2}$ units.

COORDINATE GEOMETRY

Distance (reinforcement)

The distance formula is a fundamental tool in coordinate geometry that calculates the length of the line segment joining two points in a coordinate plane. This formula is a direct application of the Pythagorean Theorem.

For two points (x_1, y_1) and (x_2, y_2) in a coordinate plane, the distance d between them is given by:

$$d = \sqrt{(x_2 - x_1)^2 + (y_2 - y_1)^2}$$

This formula can be derived by creating a right triangle with the line segment as the hypotenuse. The legs of this right triangle have lengths $|x_2 - x_1|$ and $|y_2 - y_1|$, representing the horizontal and vertical distances between the points.

By the Pythagorean Theorem, the square of the hypotenuse equals the sum of the squares of the legs:

$$d^2 = |x_2 - x_1|^2 + |y_2 - y_1|^2 = (x_2 - x_1)^2 + (y_2 - y_1)^2$$

Taking the square root of both sides yields the distance formula.

The distance formula has several important applications in coordinate geometry:

1. Verifying congruent segments: Two line segments are congruent if they have the same length.

2. Verifying collinear points: Three or more points are collinear if the sum of the distances between consecutive points equals the distance between the endpoints.

3. Calculating the perimeter of a polygon: Sum the distances between consecutive vertices, including the distance from the last vertex to the first.

4. Determining the type of triangle: By calculating the lengths of all three sides of a triangle, one can verify whether it is equilateral, isosceles, or scalene.

5. Checking if a triangle is right-angled: Using the Pythagorean Theorem, one can verify if the square of the longest side equals the sum of the squares of the other two sides.

For example, to find the distance between points $A(3, -2)$ and $B(-1, 5)$:

$$d_{AB} = \sqrt{(-1-3)^2 + \left(5-(-2)\right)^2} = \sqrt{(-4)^2 + 7^2} = \sqrt{16 + 49} = \sqrt{65} \approx 8.06$$

The distance formula extends naturally to three-dimensional space. For two points (x_1, y_1, z_1) and (x_2, y_2, z_2) three-dimensional space, the distance is:

$$d = \sqrt{(x_2 - x_1)^2 + (y_2 - y_1)^2 + (z_2 - z_1)^2}$$

This extension follows from the three-dimensional version of the Pythagorean Theorem.

Midpoint (reinforcement)

The midpoint formula identifies the coordinates of the point that lies exactly halfway between two given points in a coordinate plane. This formula is based on the concept of averaging the coordinates of the two endpoints.

For two points (x_1, y_1) and (x_2, y_2) in a coordinate plane, the midpoint M has coordinates:

$$M = \left(\frac{x_1 + x_2}{2}, \frac{y_1 + y_2}{2} \right)$$

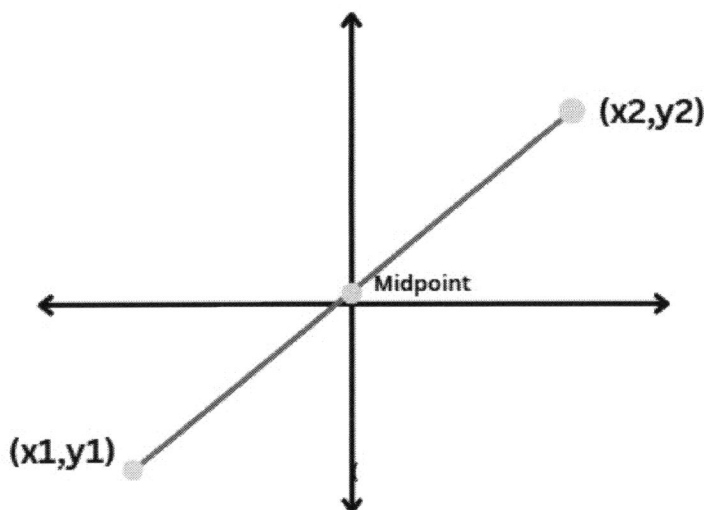

Coordinate plane showing two points and their midpoint

The midpoint formula can be derived using the concept of the arithmetic mean or by applying the distance formula to show that the midpoint is equidistant from both endpoints.

The midpoint formula has several important applications in coordinate geometry:

1. Finding the center of a line segment: The midpoint of a line segment is its center.

2. Verifying that a point is the midpoint of a segment: A point M is the midpoint of segment AB if and only if M's coordinates satisfy the midpoint formula for points A and B.

3. Locating the centroid of a triangle: The centroid can be found by calculating the coordinates of the point of intersection of the medians, or directly by averaging the coordinates of the three vertices.

4. Identifying special points in quadrilaterals: The point where the diagonals of a parallelogram intersect is the midpoint of each diagonal.

5. Creating coordinate proofs: The midpoint formula can be used to prove properties of geometric figures, such as showing that the figure formed by connecting the midpoints of consecutive sides of a quadrilateral is a parallelogram.

For example, to find the midpoint of the line segment with endpoints $C(-6, 8)$ and $D(10, -4)$:

$$M = \left(\frac{-6 + 10}{2}, \frac{8 + (-4)}{2}\right) = \left(\frac{4}{2}, \frac{4}{2}\right) = (2,2)$$

The midpoint formula extends naturally to three-dimensional space. For two points (x_1, y_1, z_1) and (x_2, y_2, z_2) in three-dimensional space, the midpoint has coordinates:

$$M = \left(\frac{x_1 + x_2}{2}, \frac{y_1 + y_2}{2}, \frac{z_1 + z_2}{2}\right)$$

An extension of the midpoint concept is the section formula, which finds the coordinates of a point that divides a line segment in a given ratio. If a point P divides the line segment from $A(x_1, y_1)$ to $B(x_2, y_2)$ in the ratio $m:n$, then the coordinates of P are:

$$P = \left(\frac{mx_2 + nx_1}{m + n}, \frac{my_2 + ny_1}{m + n}\right)$$

When $m = n = 1$, this reduces to the midpoint formula.

Slope and Its Applications

The slope of a line measures its steepness and direction, providing a quantitative way to describe how quickly a line rises or falls as it moves from left to right

For a line passing through two points (x_1, y_1) and (x_2, y_2), the slope m is defined as:

$$m = \frac{y_2 - y_1}{x_2 - x_1}$$

provided that $x_2 \neq x_1$.

This formula represents the ratio of the vertical change (rise) to the horizontal change (run) as one moves from one point to another along the line. A positive slope indicates that the line rises from left to right, while a negative slope indicates that it falls from left to right.

The slope has several special cases:

- A horizontal line has a slope of zero ($m = 0$)

- A vertical line has an undefined slope (division by zero).

- A line with a slope of 1 rises at a 45° angle to the positive x-axis.

- A line with a slope of -1 falls at a 45° angle to the positive x-axis.

The slope has numerous applications in coordinate geometry:

1. Determining parallelism: Two non-vertical lines are parallel if and only if they have the same slope. That is, if lines l_1 and l_2 have slopes m_1 and m_2 respectively, then $l_1 \parallel l_2$ if and only if $m_1 = m_2$.

2. Determining perpendicularity: Two non-vertical lines are perpendicular if and only if the product of their slopes is -1. That is, if lines l_1 and l_2 have slopes m_1 and m_2 respectively, then $l_1 \perp l_2$ if and only if $m_1 \cdot m_2 = -1$.

3. Calculating the angle between two lines: If lines with slopes m_1 and m_2 intersect at angle θ, then:

$$\tan \theta = \left| \frac{m_2 - m_1}{1 + m_1 m_2} \right|$$

4. Verifying collinearity: Three or more points are collinear if and only if any two pairs of consecutive points have the same slope.

5. Identifying special quadrilaterals: A quadrilateral is a parallelogram if and only if its opposite sides have the same slope. A quadrilateral is a rhombus if it is a parallelogram and its diagonals have slopes that are negative reciprocals of each other.

For example, to find the slope of the line through points P(3, 5) and Q(-2, -4):

$$m = \frac{-4 - 5}{-2 - 3} = \frac{-9}{-5} = \frac{9}{5} = 1.8$$

The concept of slope extends to three-dimensional space, where the direction of a line is described by its direction ratios or direction cosines.

Equations of Lines

The equation of a line in a coordinate plane can be expressed in several forms, each highlighting different aspects of the line. The two most common forms are the slope-intercept form and the point-slope form.

The slope-intercept form of a line with slope m and y-intercept b is:

$$y = mx + b$$

where m represents the slope of the line, and b represents the y-coordinate of the point where the line crosses the y-axis.

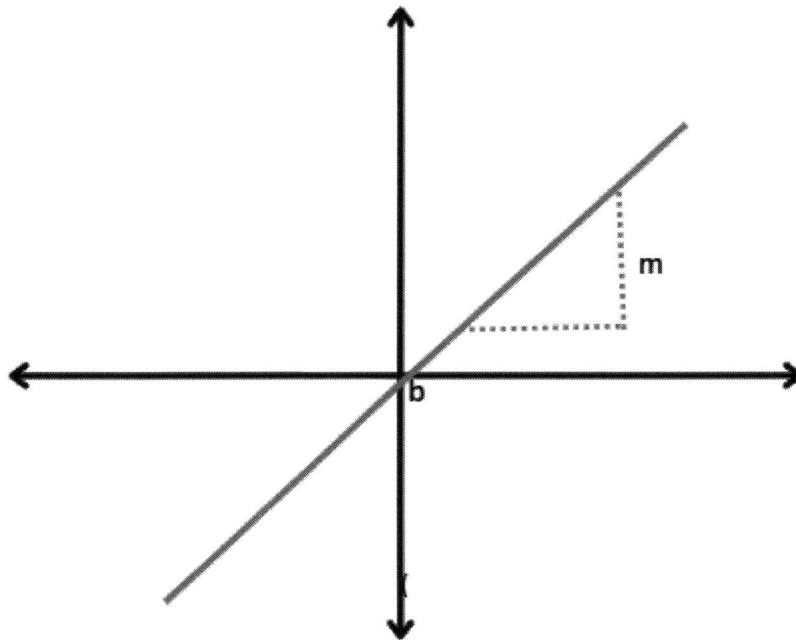

Coordinate plane showing a line with slope m and y-intercept b

This form is particularly useful for graphing lines (the slope indicates the steepness, and the y-intercept gives a starting point) and for quickly identifying the slope and y-intercept of a line from its equation.

The point-slope form of a line that passes through the point (x_1, y_1) with slope m is:

$$y - y_1 = m(x - x_1)$$

This form emphasizes that the line passes through a specific point and has a specific slope. It is especially useful when the coordinates of a point on the line are known, rather than the y-intercept.

The standard form of a line is:

$$Ax + By + C = 0$$

where A, B, and C are constants, and A and B are not both zero. This form is often used in systems of linear equations and in problems involving the distance from a point to a line.

Converting between these forms involves algebraic manipulation:

1. To convert from point-slope to slope-intercept:

   Start with $y - y_1 = m(x - x_1)$

   Expand: $y - y_1 = mx - mx_1$

   Solve for y: $y = mx - mx_1 + y_1 = mx + (y_1 - mx_1)$

   The y-intercept is $b = y_1 - mx_1$

2. To convert from slope-intercept to standard form:

   Start with $y = mx + b$

   Rearrange: $y - mx - b = 0$

   Which gives $-mx + y - b = 0$ or $Ax + By + C = 0$ where $A = -m, B = 1$, and $C = -b$

Lines can be defined by two points, a point and a slope, or other combinations of information. Given two distinct points (x_1, y_1) and (x_2, y_2), the equation of the line through these points can be found by:

1. Calculating the slope: $m = \frac{y_2 - y_1}{x_2 - x_1}$

2. Using the point-slope form with either of the given points

3. Converting to the desired form if necessary

Special cases include:

- A horizontal line has the equation $y = b$, where b is the y-coordinate of any point on the line.

- A vertical line has the equation $x = a$, where a is the x-coordinate of any point on the line.

For example, to find the equation of the line passing through the points A(3, -2) and B(-1, 6) in slope-intercept form:

First, calculate the slope: $m = \frac{6-(-2)}{-1-3} = \frac{8}{-4} = -2$

Using the point-slope form with point A(3, -2): $y - (-2) = -2(x - 3)$ $y + 2 = -2x + 6$ $y = -2x + 4$

So the equation in slope-intercept form is $y = -2x + 4$.

Parallel and Perpendicular Lines

Parallel and perpendicular lines have specific relationships in terms of their slopes, providing a powerful algebraic method for analyzing these geometric concepts in the coordinate plane.

Two distinct non-vertical lines with slopes m_1 and m_2 are parallel if and only if their slopes are equal:

$$l_1 \parallel l_2 \text{ if and only if } m_1 = m_2$$

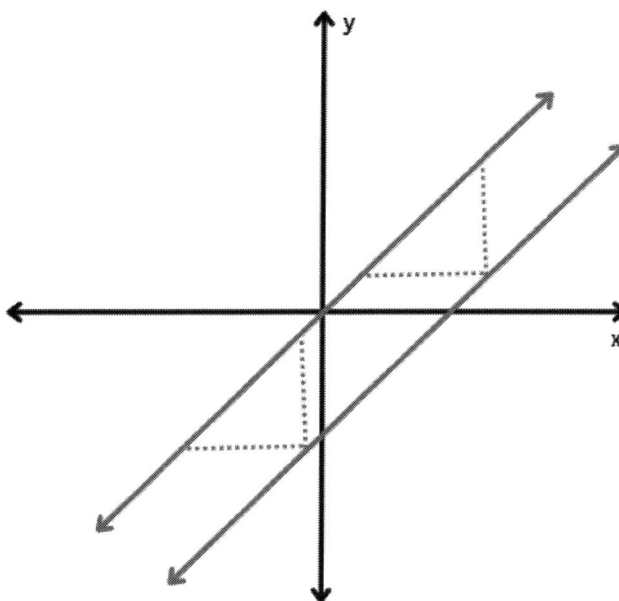

Coordinate plane showing two parallel lines with the same slope

This property follows from the definition of slope as the measure of a line's steepness. Parallel lines maintain the same steepness throughout their length, hence they must have the same slope.

Two non-vertical lines with slopes m_1 and m_2 are perpendicular if and only if their slopes are negative reciprocals of each other.

$$l_1 \perp l_2 \text{ if and only if } m_1 \cdot m_2 = -1 \text{ or } m_1 = -\frac{1}{m_2}$$

This property can be proved using the fact that two lines are perpendicular if they form a right angle (90°) at their intersection. When two lines intersect, the tangent of the angle between them is given by:

87

$$\tan \theta = \left| \frac{m_2 - m_1}{1 + m_1 m_2} \right|$$

For $\theta = 90°$, $\tan \theta$ is undefined, which occurs when the denominator $1 + m_1 m_2 = 0$, or $m_1 m_2 = -1$.

Special cases include:

- A horizontal line (with slope 0) is perpendicular to a vertical line (with undefined slope).

- Two horizontal lines are parallel to each other.

- Two vertical lines are parallel to each other.

These slope relationships have numerous applications in coordinate geometry:

1. Finding parallel or perpendicular lines: Given the equation of a line and a point not on the line, the equation of a line through the point that is parallel or perpendicular to the given line can be determined using the slope conditions.

2. Verifying geometric figures: The properties of parallel and perpendicular sides can be used to verify that a quadrilateral is a parallelogram, rectangle, rhombus, or square.

3. Constructing geometric figures: Parallel and perpendicular lines can be used to construct figures with specific properties in the coordinate plane.

4. Solving geometric problems: The relationships between parallel and perpendicular lines can be used to solve problems involving distances, angles, and areas.

For example, to find the equation of a line that passes through the point (2, -3) and is perpendicular to the line $y = 2x + 5$:

First, identify the slope of the given line: $m_1 = 2$. The slope of a perpendicular line is $m_2 = -\frac{1}{m_1} = -\frac{1}{2}$.

Using the point-slope form with the point (2, -3) and slope $-\frac{1}{2}$: $y - (-3) = -\frac{1}{2}(x - 2)$ $y + 3 = -\frac{1}{2}x + 1$ $y = -\frac{1}{2}x - 2$

So the equation of the perpendicular line is $y = -\frac{1}{2}x - 2$.

Coordinate Proofs

Coordinate proofs leverage the power of analytic geometry to establish geometric theorems and properties. By placing geometric figures in a coordinate system, algebraic methods can be applied to prove geometric relationships.

The general approach to coordinate proofs involves:

1. Placing the geometric figure in a coordinate system in a convenient way
2. Representing the geometric elements (points, lines, angles, distances) using coordinates and algebraic equations
3. Applying algebraic techniques to verify the desired geometric properties

Strategic placement of figures in the coordinate system can greatly simplify the algebraic work. Common strategies include:

*   Placing a vertex at the origin
*   Positioning a side along an axis
*   Aligning a figure with the coordinate axes
*   Using symmetry to simplify coordinates

For example, to prove that the diagonals of a parallelogram bisect each other, one could place the parallelogram in the coordinate system with vertices at $(0,0)$, $(a,0)$, (b,c), and $(a+b,c)$, where a, b, and c are positive real numbers. This ensures that the figure is a parallelogram because opposite sides have the same slope:

The slope of the side from $(0,0)$ to $(a,0)$ is 0. The slope of the side from (b,c) to $(a+b,c)$ is also 0. The slope of the side from $(0,0)$ to (b,c) is c/b. The slope of the side from $(a,0)$ to $(a+b,c)$ is also c/b.

The diagonals are from $(0,0)$ to $(a+b,c)$ and from $(a,0)$ to (b,c). Their midpoints are: $M_1 = \left(\frac{0+(a+b)}{2}, \frac{0+c}{2}\right) = \left(\frac{a+b}{2}, \frac{c}{2}\right)$, $M_2 = \left(\frac{a+b}{2}, \frac{0+c}{2}\right) = \left(\frac{a+b}{2}, \frac{c}{2}\right)$

Since $M_1 = M_2$, the diagonals bisect each other.

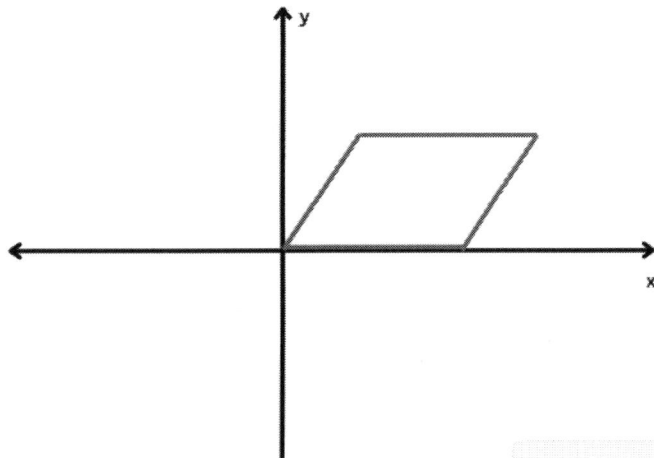

Coordinate plane showing a parallelogram

Coordinate proofs can be used to establish various geometric theorems:

1. The midsegment theorem: The segment connecting the midpoints of two sides of a triangle is parallel to the third side and half its length.

2. The centroid of a triangle: The centroid (the point of intersection of the medians) divides each median in a 2:1 ratio.

3. Properties of special quadrilaterals: For example, proving that the diagonals of a rhombus are perpendicular, or that the diagonals of a rectangle are congruent.

4. The concurrency of special lines in a triangle: For example, proving that the perpendicular bisectors of the sides of a triangle are concurrent.

For instance, to prove that the midpoint of the hypotenuse of a right triangle is equidistant from all three vertices, one could place the right triangle in the coordinate system with vertices at $(0,0)$, $(a,0)$, and $(0,b)$, where a and b are positive real numbers. The midpoint of the hypotenuse is: $M = \left(\frac{a+0}{2}, \frac{0+b}{2}\right) = \left(\frac{a}{2}, \frac{b}{2}\right)$

The distances from M to each vertex are: $d_{M,(0,0)} = \sqrt{\left(\frac{a}{2}\right)^2 + \left(\frac{b}{2}\right)^2} = \frac{\sqrt{a^2+b^2}}{2}$ $d_{M,(a,0)} = \sqrt{\left(\frac{a}{2}-a\right)^2 + \left(\frac{b}{2}-0\right)^2} =$

$\sqrt{\left(-\frac{a}{2}\right)^2 + \left(\frac{b}{2}\right)^2} = \frac{\sqrt{a^2+b^2}}{2}$ $d_{M,(0,b)} = \sqrt{\left(\frac{a}{2}-0\right)^2 + \left(\frac{b}{2}-b\right)^2} = \sqrt{\left(\frac{a}{2}\right)^2 + \left(-\frac{b}{2}\right)^2} = \frac{\sqrt{a^2+b^2}}{2}$

Since all three distances are equal, the midpoint of the hypotenuse is equidistant from all three vertices.

Conic Sections

Conic sections are curves obtained by intersecting a cone with a plane. Depending on the angle of the plane relative to the cone, four types of curves can be formed: circles, ellipses, parabolas, and hyperbolas. Each of these curves has a specific equation in the coordinate plane.

A circle is the set of all points in a plane that are equidistant from a fixed point called the center. The standard form of the equation of a circle with center (h, k) and radius r is:

$$(x - h)^2 + (y - k)^2 = r^2$$

This equation directly expresses the distance formula: every point (x, y) on the circle is at distance r from the center (h, k).

A parabola is the set of all points in a plane that are equidistant from a fixed point (the focus) and a fixed line (the directrix). The standard form of the equation of a parabola with vertex at (h, k) is:

$$(x - h)^2 = 4p(y - k) \text{ (opening vertically)}$$

or

$$(y - k)^2 = 4p(x - h) \text{ (opening horizontally)}$$

where |p| is the distance from the vertex to the focus, and the sign of p determines the direction of opening.

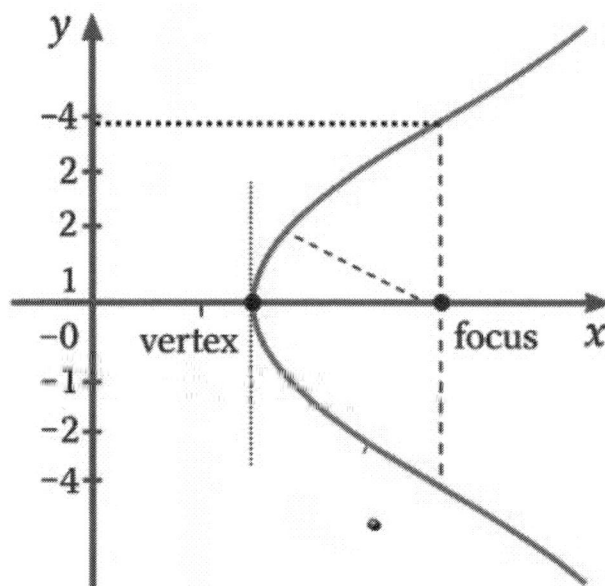

An ellipse is the set of all points in a plane such that the sum of the distances from any point on the ellipse to two fixed points (the foci) is constant. The standard form of the equation of an ellipse with center at (h, k) is:

$$\frac{(x-h)^2}{a^2} + \frac{(y-k)^2}{b^2} = 1$$

where a and b are the lengths of the semi-major and semi-minor axes, respectively.

If $a > b$, the ellipse is elongated along the x-axis, and the foci are at $(h \pm c, k)$, where $c = \sqrt{a^2 - b^2}$. If $b > a$, the ellipse is elongated along the y-axis, and the foci are at $(h, k \pm c)$, where $c = \sqrt{b^2 - a^2}$. A hyperbola is the set of all points in a plane such that the absolute difference of the distances from any point on the hyperbola to two fixed points (the foci) is constant. The standard form of the equation of a hyperbola with center at (h, k) is:

$$\frac{(x-h)^2}{a^2} - \frac{(y-k)^2}{b^2} = 1 \text{ (opening horizontally)}$$

or

$$\frac{(y-k)^2}{a^2} - \frac{(x-h)^2}{b^2} = 1 \text{ (opening vertically)}$$

For a hyperbola opening horizontally, the foci are at $(h \pm c, k)$, where $c = \sqrt{a^2 + b^2}$, and the asymptotes have equations $y - k = \pm \frac{b}{a}(x - h)$.

For a hyperbola opening vertically, the foci are at $(h, k \pm c)$, where $c = \sqrt{a^2 + b^2}$, and the asymptotes have equations $y - k = \pm \frac{a}{b}(x - h)$.

Each conic section can also be expressed in terms of its eccentricity e:

- For a circle, e = 0

- For an ellipse, 0 < e < 1

- For a parabola, e = 1

- For a hyperbola, e > 1

These curves have numerous applications:

- Circles are used in measuring distances and in circular motion.

- Parabolas appear in projectile motion, satellite dishes, and headlight reflectors.

- Ellipses are used in planetary orbits, whispering galleries, and medical lithotripsy.

- Hyperbolas are used in navigation systems (LORAN), cooling towers, and atomic-particle accelerators.

Practice Problems

1. Find the distance between the points $(-3, 5)$ and $(4, -2)$.

2. Determine the midpoint of the line segment with endpoints $A(6, -3)$ and $B(-2, 7)$.

3. Find the equation of the line passing through the point $(3, -1)$ and perpendicular to the line $2x - 3y + 6 = 0$.

Solutions

1. To find the distance between points (-3, 5) and (4, -2):

Using the distance formula:

$$d = \sqrt{(x_2 - x_1)^2 + (y_2 - y_1)^2}, d = \sqrt{\left(4 - (-3)\right)^2 + (-2 - 5)^2},$$

$$d = \sqrt{7^2 + (-7)^2}, d = \sqrt{49 + 49},$$

$$d = \sqrt{98}, d = 7\sqrt{2} \approx 9.90$$

The distance between the points is $7\sqrt{2}$ units.

2. To determine the midpoint of the line segment with endpoints $A(6, -3)$ and $B(-2, 7)$:

Using the midpoint formula:

$$M = \left(\frac{x_1 + x_2}{2}, \frac{y_1 + y_2}{2}\right), M = \left(\frac{6 + (-2)}{2}, \frac{-3 + 7}{2}\right),$$

$$M = \left(\frac{4}{2}, \frac{4}{2}\right), M = (2,2)$$

The midpoint of the line segment is $(2, 2)$.

3. To find the equation of the line passing through the point $(3, -1)$ and perpendicular to the line $2x - 3y + 6 = 0$:

First, rearrange the given line equation to slope-intercept form: $2x - 3y + 6 = -3y = -2x - y = \frac{2}{3}x + 2$

The slope of the given line is $\frac{2}{3}$.

For perpendicular lines, the slopes multiply to give -1. So the slope of the perpendicular line is: $m_{perpendicular} = -\frac{1}{m_{given}} = -\frac{1}{\frac{2}{3}} = -\frac{3}{2}$

93

Using the point-slope form with the point (3, -1) and slope $-\frac{3}{2}$: $y - (-1) = -\frac{3}{2}(x - 3)$ $y + 1 = -\frac{3}{2}x + \frac{9}{2}$ $y = -\frac{3}{2}x + \frac{9}{2} - 1$ $y = -\frac{3}{2}x + \frac{7}{2}$.

The equation of the perpendicular line is $y = -\frac{3}{2}x + \frac{7}{2}$ or in standard form: $3x + 2y - 7 = 0$.

GEOMETRIC MEASUREMENT AND DIMENSION

Perimeter and Circumference

Perimeter and circumference are measurements of the distance around the boundary of two-dimensional figures. The perimeter of a polygon is the sum of the lengths of all its sides. For a polygon with n sides, where the length of each side is denoted by $s_1, s_2, ..., s_n$, the perimeter P is given by:

$$P = s_1 + s_2 + \cdots + s_n$$

For regular polygons, where all sides have the same length s, the perimeter simplifies to:

$$P = n \times s$$

where n is the number of sides.

For specific polygons, the perimeter formulas are:

1. Rectangle with length l and width w: $P = 2l + 2w$
2. Square with side length s: $P = 4s$
3. Triangle with sides a, b, and c: $P = a + b + c$

The circumference of a circle is the distance around its boundary. For a circle with radius r or diameter d, the circumference C is given by:

$$C = 2\pi r = \pi d$$

where π is approximately 3.14159.

Perimeter and circumference calculations often involve composite figures, which are combinations of basic shapes. For such figures, the total perimeter is calculated by identifying the boundary segments and summing their lengths, being careful not to count shared segments twice.

When working with perimeter and circumference in coordinate geometry, the distance formula can be applied to find the length of each segment connecting consecutive vertices:

$$d = \sqrt{(x_2 - x_1)^2 + (y_2 - y_1)^2}$$

In practical applications, perimeter and circumference measurements are used to determine quantities such as:

- The amount of fencing needed for an enclosure

- The length of a border around a decorative element

- The distance traveled in one complete rotation of a wheel

- The amount of trim needed for a circular window

Area of Polygons and Circles

Area measures the two-dimensional space enclosed by a figure. The area of a rectangle with length l and width w is given by:

$$A = l \times w$$

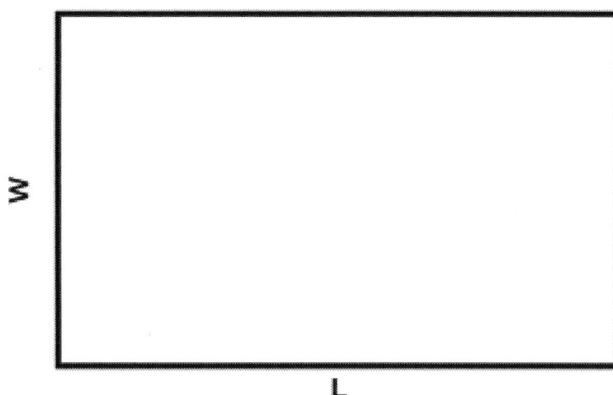

A Rectangle

The area of a square with side length s is a special case of the rectangle formula:

$$A = s^2$$

The area of a triangle can be calculated using several equivalent formulas:

1. Using base b and height h: $A = \frac{1}{2} \times b \times h$

2. Using the semi-perimeter formula (Heron's formula) with sides a, b, and c: $A = \sqrt{s(s - a)(s - b)(s - c)}$

   where s is the semi-perimeter: $s = \frac{a+b+c}{2}$

3. Using coordinates (for a triangle with vertices at (x_1, y_1), (x_2, y_2), and (x_3, y_3): $A = \frac{1}{2}|x_1(y_2 - y_3) + x_2(y_3 - y_1) + x_3(y_1 - y_2)|$

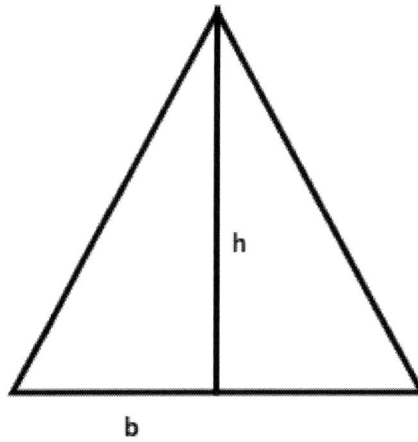

A Triangle

The area of a parallelogram with base b and height h is:

$$A = b \times h$$

where h is the perpendicular height to the base.

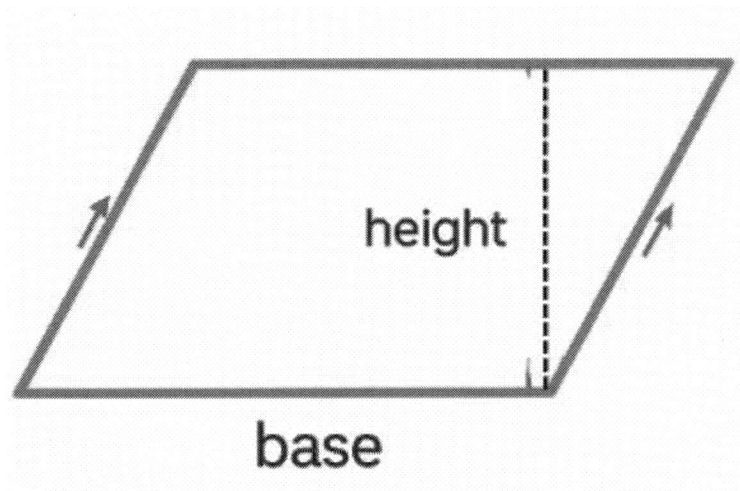

A Parallelogram

The area of a trapezoid with parallel sides a and c, and height h is:

$$A = \frac{1}{2} \times h \times (a + c)$$

97

where a and c are the lengths of the parallel sides.

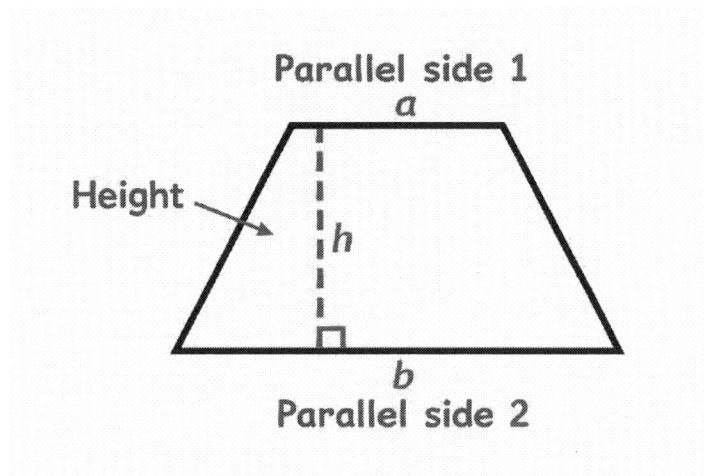

A Trapezoid

The area of a rhombus can be calculated using its diagonals d_1 and d_2:

$$A = \frac{1}{2} \times d_1 \times d_2$$

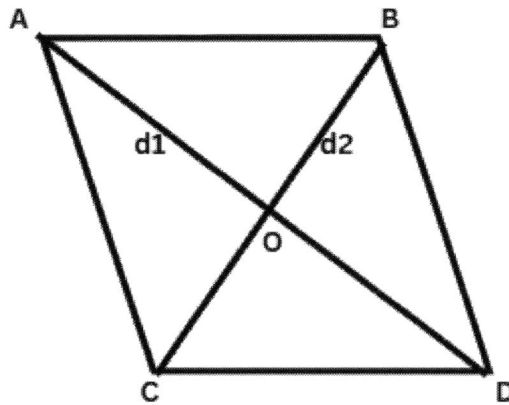

A Rhombus

The area of a regular polygon with n sides, side length s, and apothem a (the perpendicular distance from the center to any side) is:

$$A = \frac{1}{2} \times n \times s \times a$$

This can also be written as:

$$A = \frac{1}{2} \times P \times a$$

where P is the perimeter.

The area of a circle with radius r is:

$$A = \pi r^2$$

For sectors of a circle, where the central angle is θ (in radians), the area is:

$$A = \frac{1}{2} \times r^2 \times \theta$$

If the angle is measured in degrees, the formula becomes:

$$A = \frac{\theta}{360°} \times \pi r^2$$

Area calculations for composite figures involve dividing the figure into simpler shapes, calculating the area of each, and then summing these areas. If the composite figure has holes, the areas of the holes are subtracted from the total area.

In coordinate geometry, the Shoelace formula (also known as the Surveyor's formula) can be used to find the area of any polygon given the coordinates of its vertices $(x_1, y_1), (x_2, y_2), \ldots, (x_n, y_n)$:

$$A = \frac{1}{2} \left| \sum_{i=1}^{n} x_i (y_{i+1} - y_{i-1}) \right|$$

where $y_{n+1} = y_1$ and $y_0 = y_n$.

Surface Area of 3D Figures

Surface area is the total area of all the surfaces that enclose a three-dimensional object. Calculating surface area is crucial for applications such as determining the amount of material needed to cover an object, the amount of paint required for a structure, or the rate of heat transfer through a surface.

The surface area of a rectangular prism (or box) with length l, width w, and height h is the sum of the areas of its six rectangular faces:

$$SA = 2lw + 2lh + 2wh$$

This can also be written as:

$$SA = 2(lw + lh + wh)$$

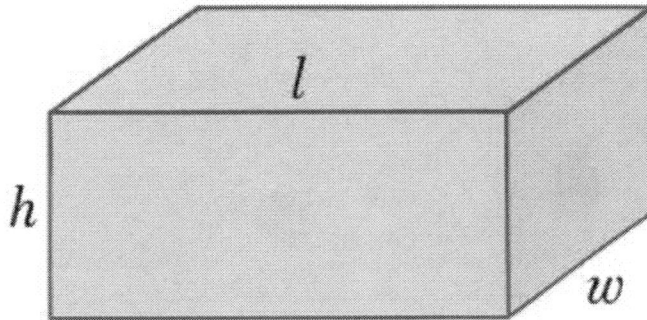

A Rectangular prism

The surface area of a cube with side length s is a special case of the rectangular prism formula:

$$SA = 6s^2$$

The surface area of a cylinder with radius r and height h consists of two circular bases and a rectangular lateral surface:

$$SA = 2\pi r^2 + 2\pi rh = 2\pi r(r + h)$$

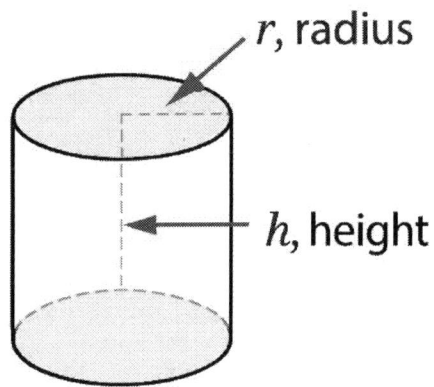

Cylinder

The surface area of a cone with radius r and slant height l consists of a circular base and a curved lateral surface:

$$SA = \pi r^2 + \pi rl$$

where the slant height l is related to the radius r and the height h by the Pythagorean Theorem:

$$l = \sqrt{r^2 + h^2}$$

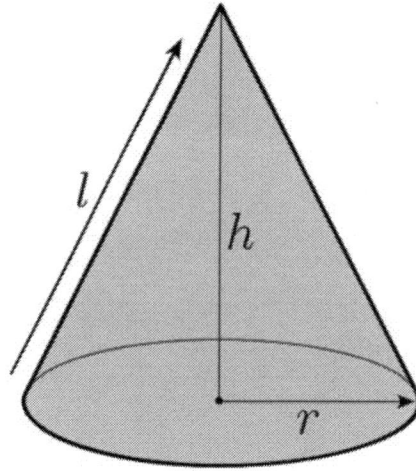

Cone

The surface area of a sphere with radius r is:

$$SA = 4\pi r^2$$

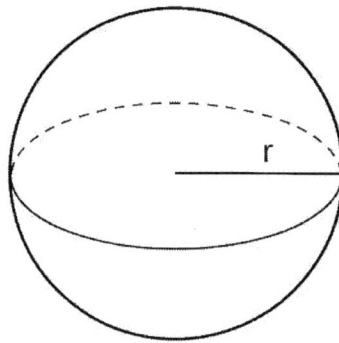

Sphere

The surface area of a rectangular pyramid with a base of length a and width b, and with triangular faces having slant heights l_1 and l_2, is:

$$SA = ab + a \times l_1 + b \times l_2$$

where l_1 and l_2 are the slant heights of the triangular faces.

101

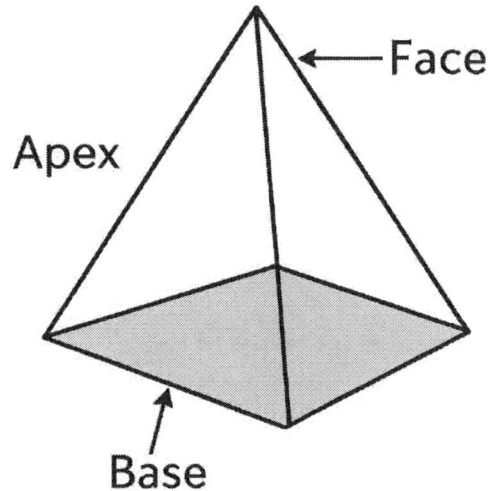

Rectangular pyramid

For a regular pyramid with an n-sided regular polygonal base of side length s, height h, and slant height l, the surface area is:

$$SA = \text{Area of base} + \frac{1}{2} \times \text{Perimeter of base} \times l$$

$$SA = \frac{n \times s^2}{4 \times \tan\left(\frac{\pi}{n}\right)} + \frac{n \times s \times l}{2}$$

The surface area of a triangular prism with triangular faces of area A_1 and A_2 and rectangular faces with areas determined by the perimeter p of the base and height h is:

$$SA = A_1 + A_2 + p \times h$$

For composite three-dimensional figures, the total surface area is calculated by identifying all the exposed surfaces, determining their areas, and summing these areas. Shared faces between component shapes are not included in the total surface area.

In real-world applications, surface area calculations may need to account for irregular shapes, which can be approximated by dividing the shape into simpler components or by using integration techniques for curved surfaces.

Volume

Volume measures the three-dimensional space occupied by a solid.

The volume of a rectangular prism (or box) with length l, width w, and height h is:

$$V = l \times w \times h$$

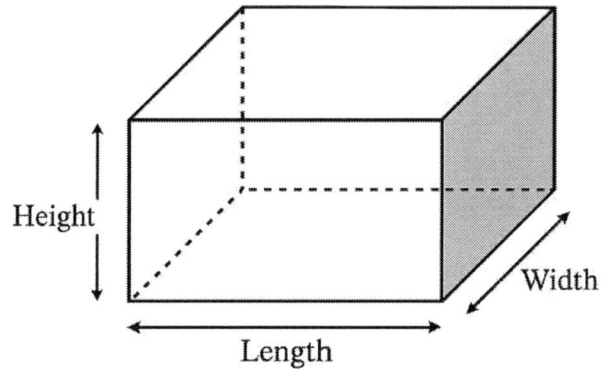

The volume of a cube with side length s is a special case of the rectangular prism formula:

$$V = s^3$$

The volume of a cylinder with radius r and height h is:

$$V = \pi r^2 h$$

The volume of a cone with radius r and height h is:

$$V = \frac{1}{3}\pi r^2 h$$

The volume of a sphere with radius r is:

$$V = \frac{4}{3}\pi r^3$$

The volume of a pyramid with base area B and height h is:

$$V = \frac{1}{3}Bh$$

For specific types of pyramids:

- Rectangular pyramid with base length a, width b, and height h: $V = \frac{1}{3} \times a \times b \times h$

- Triangular pyramid (tetrahedron) with triangular base of area A and height h: $V = \frac{1}{3} \times A \times h$

- Regular pyramid with an n-sided regular polygonal base of side length s and height h: $V = \frac{n \times s^2 \times h}{12 \times \tan\left(\frac{\pi}{n}\right)}$

The volume of a prism with base area B and height h is:

$$V = B \times h$$

For specific types of prisms:

- Triangular prism with triangular base of area A and length l: $V = A \times l$

- Regular prism with an n-sided regular polygonal base of side length s, apothem a, and height h: $V = \frac{1}{2} \times n \times s \times a \times h$

The volume of a frustum (a pyramid or cone with the top cut off parallel to the base) can be calculated using the formula:

$$V = \frac{1}{3}h\left(B_1 + B_2 + \sqrt{B_1 B_2}\right)$$

where B_1 and B_2 are the areas of the two parallel faces, and h is the height of the frustum.

For composite three-dimensional figures, the total volume is the sum of the volumes of the component parts. If one solid has a cavity occupied by another solid, the volume of the cavity is subtracted from the total.

Volume calculations have numerous applications:
- Determining the capacity of containers and storage spaces
- Calculating the weight of an object based on its volume and density
- Estimating materials needed for construction or manufacturing
- Analyzing fluid flow and displacement
- Designing efficient packaging and shipping solutions

In coordinate geometry, the volume of a solid can be calculated using triple integrals or by applying formulas to shapes defined by coordinates in three-dimensional space.

Geometric Modeling and Applications

Geometric modeling applies mathematical principles to represent and analyze real-world objects and phenomena. This field bridges theoretical geometry and practical applications, enabling problem-solving across diverse disciplines.

Geometric models use mathematical equations and geometric principles to describe physical objects and spatial relationships. These models can be:

- Two-dimensional (2D): Representing objects as points, lines, and shapes in a plane
- Three-dimensional (3D): Representing objects as solids in space
- Higher-dimensional: Representing complex systems with multiple parameters

Geometric modeling techniques include:

1. Boundary representation (B-rep): Defining objects by their boundaries (vertices, edges, and faces)
2. Constructive solid geometry (CSG): Building complex objects by combining simple primitives through operations like union, intersection, and difference
3. Parametric modeling: Defining objects using parameters that can be adjusted to change the object's dimensions or shape
4. Feature-based modeling: Creating models by adding, removing, or modifying features of a base object

Applications of geometric modeling span numerous fields:

In architecture and engineering, geometric models help design structures, analyze structural integrity, and visualize projects before construction. For example, the cross-sectional area and moment of inertia of a beam can be calculated using geometric formulas:

$$I = \frac{bh^3}{12}$$

where I is the moment of inertia, b is the width, and h is the height of a rectangular cross-section.

In manufacturing, geometric models guide production processes, determine material requirements, and optimize product design. The volume of material needed can be calculated using appropriate formulas:

$$V_{material} = V_{outer} - V_{inner}$$

where V_{outer} is the volume of the outer boundary and V_{inner} is the volume of any hollow spaces.

In computer graphics and animation, geometric models create realistic representations of objects and scenes. These models use vertices, edges, and faces to define shapes, with transformations applied to create animation:

$$P' = T \times R \times S \times P$$

where P is the original position, T is the translation matrix, R is the rotation matrix, S is the scaling matrix, and P' is the transformed position.

In scientific research, geometric models simulate natural phenomena, visualize data, and test hypotheses. For example, the surface area-to-volume ratio of a cell affects its ability to exchange materials with its environment:

$$\frac{SA}{V} = \frac{4\pi r^2}{\frac{4}{3}\pi r^3} = \frac{3}{r}$$

This ratio increases as size decreases, explaining why smaller cells are more efficient at material exchange.

In robotics and autonomous systems, geometric models enable spatial reasoning, path planning, and obstacle avoidance. The configuration space represents all possible positions of a robot, with obstacles mapped as forbidden regions.

Optimization problems often employ geometric modeling to find optimal solutions. For example, finding the dimensions of a rectangular box with minimum surface area, given a fixed volume V:

$$SA = 2(xy + xz + yz)V = xyz$$

Using calculus and constraints, the optimal solution is a cube with side length $\sqrt[3]{V}$.

Scaling relationships in geometric modeling follow important principles:

- When linear dimensions scale by a factor k, area scales by k^2, and volume scales by k^3

- This relationship explains why larger animals need proportionally stronger bones and why heat dissipation becomes more challenging with increased size

Practice Problems

1. A rectangular garden has a length of 12 feet and a width of 8 feet. Find the perimeter and area of the garden.

2. A circular pizza has a diameter of 14 inches. Find the circumference and area of the pizza. Use $\pi \approx 3.14$.

3. A triangular prism has a base in the shape of a right triangle with legs measuring 6 cm and 8 cm. If the height of the prism is 10 cm, find the surface area and volume of the prism.

4. A cone has a radius of 5 meters and a height of 12 meters. Calculate its volume and lateral surface area (excluding the base).

5. A water tank is in the shape of a cylinder with radius 4 feet and height 6 feet. How many gallons of water can the tank hold? (1 cubic foot ≈ 7.48 gallons)

Solutions

1. For a rectangular garden with length 12 feet and width 8 feet:

$$Perimeter = 2(length) + 2(width) = 2(12\,ft) + 2(8\,ft) = 24\,ft + 16\,ft = 40\,ft$$

$$Area = length \times width = 12\,ft \times 8\,ft = 96\,ft^2$$

The garden has a perimeter of 40 feet and an area of 96 square feet.

2. For a circular pizza with diameter 14 inches:

$$Radius = diameter \div 2 = 14\,in \div 2 = 7\,in$$

$$Circumference = 2\pi r = 2\pi(7\,in) = 14\pi\,in \approx 14(3.14)\,in \approx 43.96\,in$$

$$Area = \pi r^2 = \pi(7\,in)^2 = 49\pi\,in^2 \approx 49(3.14)\,in^2 \approx 153.86\,in^2$$

The pizza has a circumference of approximately 43.96 inches and an area of approximately 153.86 square inches.

3. For a triangular prism with a right triangular base having legs 6 cm and 8 cm, and height 10 cm:

First, find the hypotenuse of the base using the Pythagorean theorem:

$$c^2 = 6^2 + 8^2 = 36 + 64 = 100\ c = 10\,cm$$

$$Area\ of\ triangular\ base = (1/2) \times 6\,cm \times 8\,cm = 24\,cm^2$$

$$Surface\,area = 2(area\,of\,base) + (perimeter\,of\,base \times height)$$
$$= 2(24cm^2) + (6cm + 8cm + 10cm) \times 10cm$$

$$= 48\,cm^2 + 24\,cm \times 10\,cm = 48\,cm^2 + 240\,cm^2 = 288\,cm^2$$

$$Volume = area\ of\ base \times height = 24\,cm^2 \times 10\,cm = 240\,cm^3$$

The triangular prism has a surface area of 288 square centimeters and a volume of 240 cubic centimeters.

4. For a cone with radius 5 meters and height 12 meters:

First, calculate the slant height using the Pythagorean theorem:

$$l = \sqrt{(r^2 + h^2)} = \sqrt{(5^2 + 12^2)} - \sqrt{(25 + 144)} = \sqrt{169} = 13m$$

$$Volume = (1/3) \times \pi \times r^2 \times h = (1/3) \times \pi \times 5^2 \times 12 = (1/3) \times \pi \times 25 \times 12 = \pi \times 100$$
$$= 100\pi\,m^3 \approx 314.16\,m^3$$

$$Lateral\ surface\ area = \pi \times r \times l = \pi \times 5 \times 13 = 65\pi\,m^2 \approx 204.2\,m^2$$

The cone has a volume of approximately 314.16 cubic meters and a lateral surface area of approximately 204.2 square meters.

5. For a cylindrical water tank with radius 4 feet and height 6 feet:

$$Volume\ in\ cubic\ feet\ =\ \pi \times r^2 \times h\ =\ \pi \times 4^2 \times 6\ =\ \pi \times 16 \times 6\ =\ 96\pi\ ft^3\ \approx\ 96(3.14)\ ft^3$$
$$\approx\ 301.44\ ft^3$$

$$Volume\ in\ gallons\ =\ 301.44\ ft^3 \times 7.48\ gallons/ft^3\ \approx\ 2,254.77\ gallons$$

The water tank can hold approximately 2,255 gallons of water. $= 48\ cm^2 + 24\ cm \times 10\ cm$

PRACTICE QUESTIONS

Foundations of Geometry

1. Which of the following is considered an undefined term in geometry? A) Circle B) Line C) Triangle D) Angle

2. What is the distance between points P(-2, 5) and Q(4, -3)? A) 10 B) 6√2 C) 2√41 D) 10√2

3. Which of the following best describes a ray? A) A part of a line with one endpoint and extending infinitely in one direction B) A part of a line with two endpoints C) A straight path with no endpoints D) A curved path with one endpoint

4. What is the midpoint of the line segment with endpoints A(3, -7) and B(-5, 9)? A) (-1, 1) B) (-2, 1) C) (4, -8) D) (-4, 8)

5. In a coordinate plane, which of the following conditions indicates that points A, B, and C are collinear? A) The distance from A to C equals the distance from A to B plus the distance from B to C B) The slopes of lines AB and BC are different C) The three points form a triangle with positive area D) The angles formed by the three points sum to 180°

6. Two angles that share a common vertex and a common side are called: A) Vertical angles B) Complementary angles C) Adjacent angles D) Supplementary angles

7. If lines l and m are parallel, and line t is a transversal, which of the following is true? A) Corresponding angles are supplementary B) Alternate interior angles are congruent C) Consecutive interior angles are congruent D) Vertical angles are supplementary

8. In the figure below, lines l and m are parallel, and line t is a transversal. If angle 1 measures 65°, what is the measure of angle 5? A) 25° B) 65° C) 115° D) 155°

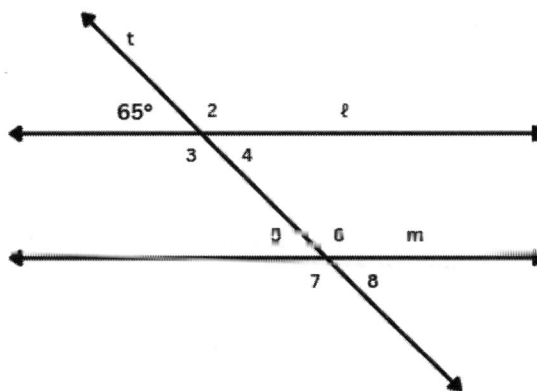

9. What is the distance from point P(3, 4) to the line $3x + 4y = 12$? A) 0 B) 1 C) 2 D) 3

10. If A(-1, 2), B(3, 6), and C(5, 4) are vertices of a triangle, what is the length of side AC? A) $2\sqrt{5}$ B) 6 C) $2\sqrt{10}$ D) 8

11. Two angles are complementary if their measures sum to: A) 90° B) 180° C) 270° D) 360°

12. The midpoint of a line segment: A) Divides the segment into two congruent parts B) Is twice as far from one endpoint as from the other C) Is always the origin in a coordinate plane D) Creates two segments with a ratio of 2:1

13. Which of the following is not a postulate of Euclidean geometry? A) Through any two points, exactly one line can be drawn B) All right angles are congruent C) Through a point not on a line, exactly two lines can be drawn parallel to the given line D) A circle can be drawn with any center and any radius

14. If lines l and m intersect, the vertically opposite angles are: A) Congruent B) Supplementary C) Complementary D) Neither congruent nor supplementary

15. What is the coordinate of the point that divides the line segment from A(2, -3) to B(8, 9) in the ratio 2:1 (starting from A)? A) (4, 1) B) (6, 5) C) (5, 3) D) (4, 3)

16. In a geometric construction, which tool is used to measure angles? A) Ruler B) Compass C) Protractor D) Straightedge

17. Which of the following constructions cannot be performed using only a compass and straightedge? A) Bisecting an angle B) Constructing a line perpendicular to a given line through a point on the line C) Trisecting an arbitrary angle D) Constructing a line parallel to a given line through a point not on the line

18. Which quadrant contains the point (-5, 7)? A) I B) II C) III D) IV

19. What is the slope of the line passing through the points (3, -1) and (7, 5)? A) 1.5 B) 2/3 C) 3/2 D) 2

20. What are the coordinates of the point that is 3 units to the right and 4 units above the origin? A) (3, 4) B) (4, 3) C) (-3, 4) D) (3, -4)

21. Two lines are perpendicular if and only if: A) Their slopes are equal B) The product of their slopes is -1 C) The sum of their slopes is 0 D) The product of their slopes is 1

22. The distance from point P(2, 3) to the origin is: A) 5 B) $\sqrt{5}$ C) $\sqrt{13}$ D) 13

23. What construction does the figure below depict shows the perpendicular line construction? A) perpendicular line construction B) Bisection of an angle C) Incomplete construction of a triangle D) Construction of a tangent

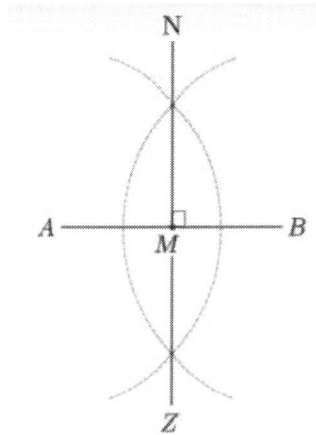

24. If the midpoint of line segment PQ is (3, -2), and P is at (5, 1), what are the coordinates of Q? A) (1, -5) B) (1, -3) C) (0, -5) D) (8, -5)

25. Which of the following statements about parallel lines is false? A) If two lines are parallel to the same line, they are parallel to each other B) Through a point not on a line, exactly one line can be drawn parallel to the given line C) Parallel lines have the same slope D) If lines are parallel, they have no points of intersection

26. What is the value of x in the equation of a circle: $(x - 3)^2 + (y + 2)^2 = 16$? A) x = 3 B) x = -3 C) x = 16 D) x = 0

27. The measure of an angle is 25°. Its complement measures: A) 25° B) 65° C) 155° D) 335°

28. In which quadrant does the point (4, -5) lie? A) I B) II C) III D) IV

29. The perpendicular bisector of a line segment: A) Passes through the midpoint of the segment B) Is parallel to the segment C) Makes a 45° angle with the segment D) Has a slope equal to the slope of the segment

30. What is the midpoint of the segment joining (a, b) and (c, d)? A) $((a + b)/2, (c + d)/2)$ B) $((a + c)/2, (b + d)/2)$ C) $((a - c)/2, (b - d)/2)$ D) $((a \times c)/2, (b \times d)/2)$

31. Which of the following points are collinear? A) (2, 3), (4, 7), (6, 11) B) (1, 2), (3, 5), (5, 7) C) (0, 0), (2, 2), (3, 4) D) (1, 1), (2, 3), (4, 4)

32. Which of the following is a basic postulate in Euclidean geometry? A) A circle can be divided into exactly 360 equal parts B) Through a point not on a line, exactly one line can be drawn parallel to the given line C) The shortest distance between two points is a straight line D) All equilateral triangles have equal angles

33. What is the distance between the points (a, b) and (a, c)? A) $|b - c|$ B) $|a - c|$ C) $\sqrt{((a - a)^2 + (b - c)^2)}$ D) $|a - b| + |a - c|$

34. Two lines that intersect at a 90° angle are: A) Perpendicular B) Parallel C) Skew D) Coincident

35. What is the slope of a horizontal line? A) 0 B) 1 C) Undefined D) Depends on the y-coordinate

36. If the endpoints of a circle's diameter are (2, 3) and (8, 9), what are the coordinates of the circle's center? A) (5, 6) B) (10, 12) C) (6, 6) D) (3, 3)

37. What is the equation of the perpendicular bisector of the line segment joining the points (2, 5) and (6, 1)? A) $y = x + 1$ B) $y = -x + 8$ C) $y = x - 3$ D) $y = -x + 9$

38. Which construction cannot be performed with compass and straightedge alone? A) Constructing an equilateral triangle B) Constructing a square C) Constructing a regular pentagon D) Constructing a regular hexagon

39. Which of the following is not an undefined term in geometry? A) Point B) Line C) Triangle D) Plane

40. A transversal intersects two lines. If corresponding angles are congruent, the lines are: A) Perpendicular B) Parallel C) Intersecting at 45° D) Forming a triangle

41. What is the equation of a circle with center (2, -3) and radius 4? A) $(x - 2)^2 + (y + 3)^2 = 16$ B) $(x + 2)^2 + (y - 3)^2 = 16$ C) $(x - 2)^2 + (y - 3)^2 = 16$ D) $(x + 2)^2 + (y + 3)^2 = 16$

42. Which of the following represents a point 3 units away from the origin? A) (3, 0) B) (0, 3) C) (3, 3) D) (1, 2)

43. What is the definition of congruent angles? A) Angles that have the same measure B) Angles that sum to 90° C) Angles that sum to 180° D) Angles that share a common vertex

44. If two lines have slopes of 2 and -1/2 respectively, what is their relationship? A) Parallel B) Perpendicular C) Same line D) Intersecting at 45°

45. The angle bisector of an angle: A) Divides the angle into two congruent angles B) Is perpendicular to one of the sides of the angle C) Creates two supplementary angles D) Creates two complementary angles

46. Which of the following is a correctly written geometric statement? A) Line AB B) Line AB C) $\overline{AB}$ D) ––>

47. If the coordinates of the midpoint of a line segment are (3, 4) and one endpoint is (1, 7), what are the coordinates of the other endpoint? A) (5, 1) B) (4, 11) C) (6, 8) D) (2, 8)

48. What is the distance between the point (3, 2) and the line y = 2? A) 0 B) 1 C) 2 D) 3

49. If angle A and angle B are complementary, and angle A measures 37°, what is the measure of angle B? A) 37° B) 53° C) 143° D) 180°

50. Which of the following represents a true statement about the coordinates of points on the x-axis? A) The x-coordinate must be zero B) The y-coordinate must be zero C) Both coordinates must be equal D) The x-coordinate must equal the y-coordinate

Logical Reasoning and Proof

1. Which of the following is the converse of the statement "If it rains, then the grass is wet"? A) If the grass is wet, then it rains B) If it does not rain, then the grass is not wet C) If the grass is not wet, then it does not rain D) If it rains, then the grass is not wet

2. A conditional statement "If p, then q" is false only when: A) p is true and q is true B) p is true and q is false C) p is false and q is true D) p is false and q is false

3. What type of reasoning moves from specific observations to general principles? A) Deductive reasoning B) Inductive reasoning C) Transitive reasoning D) Direct reasoning

4. If a conditional and its contrapositive have the same truth values, and the contrapositive has opposite truth values to the inverse, what is the relationship between the conditional and its inverse? A) They always have the same truth value B) They always have opposite truth values C) Their truth values depend on the specific statement D) They have no consistent relationship

5. Which of the following is the contrapositive of "If a triangle has three congruent sides, then it has three congruent angles"? A) If a triangle has three congruent angles, then it has three congruent sides B) If a triangle does not have three congruent sides, then it does not have three congruent angles C) If a triangle does not have three congruent angles, then it does not have three congruent sides D) If a triangle has three congruent angles, then it does not have three congruent sides

6. What type of proof shows that a statement is false by finding a specific counterexample? A) Direct proof B) Indirect proof C) Proof by contradiction D) Proof by counterexample

7. In a two-column proof, what is typically shown in the left column? A) Reasons B) Statements C) Theorems D) Definitions

8. Which of the following is a valid conclusion based on the statements "All squares are rectangles" and "All rectangles have four sides"? A) All rectangles are squares B) All squares have four sides C) All four-sided figures are rectangles D) All four-sided figures are squares

9. In the figure below, lines l and m are parallel, and line t is a transversal. If m∠1 = 115°, what is m∠8? A) 65° B) 115° C) 180° D) 245° .

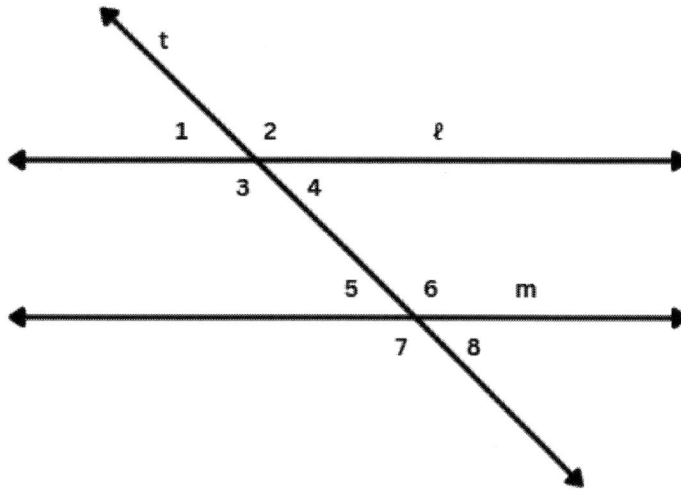

10. What is the relationship between vertical angles? A) They are complementary B) They are supplementary C) They are congruent D) They have no special relationship

11. If two lines are cut by a transversal and alternate exterior angles are congruent, what can you conclude about the two lines? A) They are perpendicular B) They are parallel C) They form a right angle D) They form complementary angles

12. Which of the following is a valid deductive argument? A) The sun has risen every day so far, so it will rise tomorrow B) Some cats are black; Felix is a cat; therefore, Felix is black C) All equilateral triangles have equal angles; Figure ABC is an equilateral triangle; therefore, Figure ABC has equal angles D) Many students who study get good grades; Maria studies; therefore, Maria will get good grades

13. What is the definition of a biconditional statement? A) A statement that is always true B) A statement that is always false C) A statement combining a conditional and its converse with "if and only if" D) A statement that has exactly two conditions

14. In a paragraph proof, what is the main difference compared to a two-column proof? A) It uses different theorems B) It presents the logical argument in continuous prose form C) It requires fewer steps D) It applies only to coordinate geometry

15. If two lines are parallel and cut by a transversal, which of the following pairs of angles are supplementary? A) Alternate interior angles B) Corresponding angles C) Consecutive interior angles D) Vertical angles

16. What is the inverse of "If an animal is a dog, then it has four legs"? A) If an animal has four legs, then it is a dog B) If an animal is not a dog, then it does not have four legs C) If an animal does not have four legs, then it is not a dog D) If an animal has four legs, then it is not a dog

17. Which type of reasoning is most commonly used in geometric proofs? A) Inductive reasoning B) Deductive reasoning C) Circular reasoning D) Analogical reasoning

18. In the statement "All right angles are congruent," what type of statement is this? A) Postulate B) Axiom C) Definition D) Theorem

19. Which of the following is the correct format for a flow proof? A) Two columns with statements and reasons B) Boxes with statements connected by arrows showing logical flow C) A paragraph with numbered statements D) A list of numbered theorems

20. If angles 1 and 2 are complementary, and angles 2 and 3 are complementary, what is the relationship between angles 1 and 3? A) They are complementary B) They are supplementary C) They are congruent D) They have no special relationship

21. When proving triangles congruent, which of the following is NOT a valid congruence criterion? A) SSS B) ASA C) AAA D) SAS

22. Which of the following is a valid syllogism? A) If it rains, the sidewalk gets wet. The sidewalk is wet. Therefore, it rained. B) All men are mortal. Socrates is mortal. Therefore, Socrates is a man. C) All mammals are animals. All dogs are mammals. Therefore, all dogs are animals. D) Some birds can fly. Penguins are birds. Therefore, penguins can fly.

23. What is a logically equivalent statement to "If I study, then I will pass the test"? A) If I don't pass the test, then I didn't study B) If I don't study, then I will pass the test C) If I pass the test, then I studied D) If I don't study, then I don't pass the test

24. In the diagram below, if $\angle A = 40°$ and $\angle C = 60°$, what is the measure of $\angle B$? A) 20° B) 40° C) 60° D) 80°

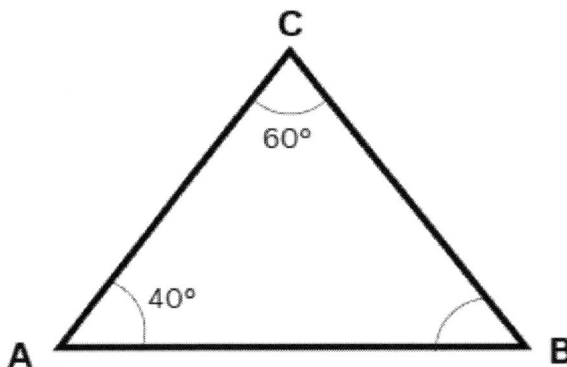

25. What is the conclusion of a proof by contradiction? A) The original statement is true B) The original statement is false C) The negation of the original statement is true D) No conclusion can be drawn

26. If $\triangle ABC \cong \triangle DEF$, which of the following is NOT necessarily true? A) AB = DE B) BC = EF C) AC = DF D) AB = BC

115

27. Which angle relationship is created when two parallel lines are cut by a transversal? A) Complementary angles B) Congruent vertical angles C) Congruent corresponding angles D) Supplementary adjacent angles

28. What is the difference between an axiom and a theorem? A) Axioms require proof, theorems are accepted without proof B) Axioms are accepted without proof, theorems require proof C) Axioms apply only to algebra, theorems apply only to geometry D) Axioms concern points and lines, theorems concern polygons

29. Which statement is logically equivalent to "If I am in Florida, then I am in the United States"? A) If I am in the United States, then I am in Florida B) If I am not in Florida, then I am not in the United States C) If I am not in the United States, then I am not in Florida D) If I am in the United States, then I am not in Florida

30. In a valid argument, if the premises are true, what can be said about the conclusion? A) It must be true B) It must be false C) It might be true or false D) Its truth value cannot be determined

31. When proving that a quadrilateral is a parallelogram, which of the following is sufficient? A) One pair of opposite sides are parallel B) Two pairs of opposite sides are congruent C) All angles are right angles D) Diagonals are perpendicular to each other

32. Which of the following is an example of the Law of Detachment? A) If $p \rightarrow q$ and $q \rightarrow r$, then $p \rightarrow r$ B) If $p \rightarrow q$ and p is true, then q is true C) If $p \rightarrow q$ and q is false, then p is false D) If $p \rightarrow q$ and $p \rightarrow r$, then $q \rightarrow r$

33. What is the negation of "All triangles have three sides"? A) Some triangles have three sides B) No triangles have three sides C) Some triangles do not have three sides D) All triangles have more than three sides

34. If "If it is a square, then it is a rectangle" is true, and "If it is a rectangle, then it has four sides" is true, what can you conclude? A) If it is a square, then it has four sides B) If it has four sides, then it is a rectangle C) If it has four sides, then it is a square D) If it is not a square, then it is not a rectangle

35. In a proof, which of the following is a valid reason for stating that two angles are congruent? A) Corresponding angles of similar triangles B) Corresponding angles of congruent triangles C) Corresponding angles of different triangles D) Corresponding angles of supplementary triangles

36. Which of the following is NOT a valid method of proof? A) Direct proof B) Indirect proof C) Proof by contradiction D) Proof by authority

37. If you know that "If p, then q" is true and "If q, then r" is false, what can you conclude about "If p, then r"? A) It must be true B) It must be false C) It could be either true or false D) Not enough information to determine

38. In a coordinate proof, what is the advantage of placing a vertex at the origin? A) It makes all coordinates positive B) It simplifies calculations by eliminating some coordinates C) It ensures that all angles are measurable D) It guarantees that the figure is in the first quadrant

39. Which of the following is a counterexample to the statement "All quadrilaterals have four congruent sides"?
A) A square B) A rhombus C) A rectangle D) A regular pentagon

40. What is the sum of the measures of the interior angles of a pentagon? A) 360° B) 540° C) 720° D) 900°

41. If angles in a triangle are in the ratio 2:3:4, what are their measures? A) 30°, 45°, 105° B) 20°, 30°, 130° C) 40°, 60°, 80° D) 60°, 60°, 60°

42. Which of the following is an example of inductive reasoning? A) All observed rectangles have four sides, so all rectangles have four sides B) All rectangles have four sides; this figure is a rectangle; therefore, this figure has four sides C) A rectangle has four sides because it is a quadrilateral D) A rectangle must have four sides by definition

43. What does the symbol "≅" represent in a geometric proof? A) Similar B) Congruent C) Equal D) Parallel

44. In a coordinate proof establishing that a quadrilateral is a parallelogram, which of the following would NOT be a valid approach? A) Showing that opposite sides have the same slope B) Showing that diagonals bisect each other C) Showing that all sides have the same length D) Showing that opposite sides are parallel

45. What is the correct statement of the transitive property? A) If $a = b$ and $b = c$, then $a = c$ B) If $a = b$, then $b = a$ C) If $a = b + c$, then $a - c = b$ D) If $a = b$, then $a + c = b + c$

46. Which of the following is NOT a postulate about lines and points in Euclidean geometry? A) Through any two points, there is exactly one line B) Through a point not on a line, there is exactly one line parallel to the given line C) Through any three points, there is exactly one circle D) Through any point, there are infinitely many lines

47. If the measure of an exterior angle of a triangle is 125° and one of the remote interior angles is 45°, what is the measure of the other remote interior angle? A) 45° B) 55° C) 80° D) 135°

48. Which of the following statements is an example of the law of contrapositive? A) If $p \rightarrow q$ is true, then $q \rightarrow p$ is true B) If $p \rightarrow q$ is true, then $\sim p \rightarrow \sim q$ is true C) If $p \rightarrow q$ is true, then $\sim q \rightarrow \sim p$ is true D) If $p \rightarrow q$ is true, then $p \wedge q$ is true

49. In a flow proof, what do the arrows represent? A) The order in which statements should be read B) The logical connections between statements C) The importance of each statement D) The types of reasons used

50. Which of the following best describes an indirect proof? A) A proof that uses coordinate geometry B) A proof that assumes the conclusion is false and derives a contradiction C) A proof that uses only definitions and no theorems D) A proof that shows the contrapositive is true.

Transformations

1. Which transformation preserves both size and shape of a figure? A) Dilation B) Translation C) Reflection D) Both B and C

2. The transformation that maps a point (x, y) to $(x + 3, y - 5)$ is a: A) Reflection B) Rotation C) Translation D) Dilation

3. A triangle with vertices at $(1, 2)$, $(3, 4)$, and $(2, 5)$ is reflected over the x-axis. What are the coordinates of the reflected vertices? A) (1, -2), (3, -4), (2, -5) B) (-1, 2), (-3, 4), (-2, 5) C) (-1, -2), (-3, -4), (-2, -5) D) (2, 1), (4, 3), (5, 2)

4. Which of the following transformations always produces a result that is congruent to the original figure? A) Translation B) Dilation C) Rotation D) Compression

5. Under reflection, what happens to the orientation of a figure? A) It is preserved B) It is reversed C) It depends on the axis of reflection D) It is rotated 90°

6. Which transformation maps point $(3, 4)$ to point $(-3, 4)$? A) Reflection over the y-axis B) Reflection over the x-axis C) Rotation of 180° around the origin D) Translation by (-6, 0)

7. Which of the following correctly represents the coordinates after a 90° counterclockwise rotation around the origin? A) (x, y) → (y, -x) B) (x, y) → (-y, x) C) (x, y) → (-x, -y) D) (x, y) → (y, x)

8. A dilation with a scale factor of 2 will: A) Double the perimeter and double the area B) Double the perimeter and quadruple the area C) Quadruple the perimeter and double the area D) Quadruple both the perimeter and the area

9. In the figure below, what single transformation maps figure A to figure B? A) Translation B) Reflection C) Rotation D) Dilation .

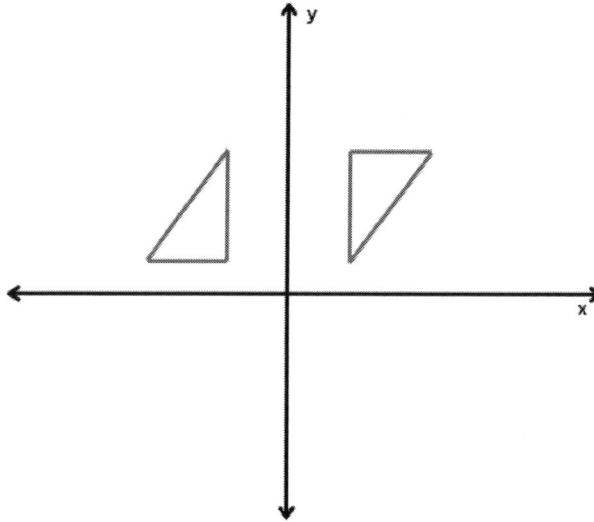

10. Which of the following is not a rigid transformation? A) Translation B) Reflection C) Rotation D) Dilation

11. Which transformation maps the point (5, 3) to (5, -3)? A) Reflection over the x-axis B) Reflection over the y-axis C) Rotation of 90° clockwise around the origin D) Translation by (0, -6)

12. Under which transformation does a line always map to a parallel line? A) Rotation B) Reflection C) Translation D) All of the above

13. A circle with center at (2, 3) and radius 4 is dilated with scale factor 3 from the origin. What are the new center and radius? A) (6, 9), radius 12 B) (6, 9), radius 7 C) (5, 6), radius 12 D) (2, 3), radius 12

14. A rotation of 180° around the origin is equivalent to: A) A reflection over the x-axis followed by a reflection over the y-axis B) A reflection over the y-axis followed by a translation C) Two 90° rotations in the same direction D) A dilation with scale factor -1

15. The composition of two reflections over parallel lines results in: A) A rotation B) A translation C) Another reflection D) A dilation

16. Which transformation does not have a fixed point? A) Reflection B) Rotation C) Translation (except identity) D) Dilation from the origin

17. A point (a, b) is reflected over the line y = x. What are the coordinates of its image? A) (-a, -b) B) (b, a) C) (-b, -a) D) (a, -b)

18. Under which transformation does the area of a figure change? A) Translation B) Rotation C) Reflection D) Dilation (except with scale factor 1)

19. What transformation maps the point (3, 5) to (-3, -5)? A) Reflection over the x-axis B) Reflection over the y-axis C) Rotation of 180° around the origin D) Translation by (-6, -10)

20. Which of the following is true about a figure with rotational symmetry of order 4? A) It can be rotated by 90° to coincide with itself B) It has exactly 4 lines of symmetry C) It must be a square D) It must have 4 sides

21. When a figure is dilated by a scale factor k from a center, the distance between any point on the figure and the center of dilation: A) Increases by k units B) Decreases by k units C) Is multiplied by k D) Is divided by k

22. Which of the following represents the rule for a reflection over the y-axis? A) $(x, y) \rightarrow (-x, y)$ B) $(x, y) \rightarrow (x, -y)$ C) $(x, y) \rightarrow (-x, -y)$ D) $(x, y) \rightarrow (y, x)$

23. The point (-2, 3) is translated by the vector $< 5, -7 >$. What are the coordinates of the image? A) (3, -4) B) (7, -10) C) (-7, 10) D) (3, 10)

24. Which transformation does not preserve distances between points? A) Translation B) Reflection C) Rotation D) Dilation (except with scale factor 1)

25. A figure has a single line of reflectional symmetry. What is the order of its rotational symmetry? A) 0 B) 1 C) 2 D) It depends on the figure

26. The composition of a reflection over the x-axis followed by a reflection over the y-axis is equivalent to: A) A reflection over the line y = x B) A rotation of 180° around the origin C) A translation D) A reflection over the origin

27. The figure shows a regular hexagon before and after a transformation. What transformation was applied? A) Reflection B) Rotation of 60° C) Dilation with scale factor 2 D) Translation

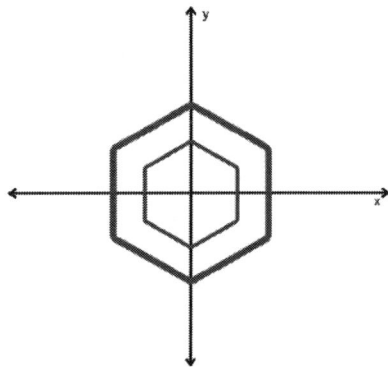

28. Which of the following types of symmetry does a circle possess? A) Reflectional symmetry only B) Rotational symmetry only C) Both reflectional and rotational symmetry D) Neither reflectional nor rotational symmetry

29. Under a dilation with scale factor k, the area of a figure is: A) Multiplied by k B) Multiplied by k^2 C) Multiplied by 2k D) Unchanged

30. A rectangle is invariant under which of the following transformations? A) Rotation of 90° around its center B) Reflection over its diagonal C) Reflection over a line through its center parallel to a side D) Dilation from its center with scale factor 2

31. Which of the following is not a property of a translation? A) Preserves orientation B) Preserves distance C) Has exactly one fixed point D) Maps parallel lines to parallel lines

32. The rule $(x, y) \rightarrow (-y, x)$ represents: A) A reflection over the line $y = x$ B) A 90° counterclockwise rotation around the origin C) A 90° clockwise rotation around the origin D) A reflection over the y-axis

33. A transformation that maps a square with area 16 to a square with area 64 is a: A) Translation B) Rotation C) Reflection D) Dilation with scale factor 2

34. Under which transformation is it possible for a figure to map onto itself? A) Translation only B) Rotation only C) Any of the rigid transformations D) None of the rigid transformations

35. The image of point (5, 2) after a 270° counterclockwise rotation around the origin is: A) (-2, 5) B) (2, -5) C) (5, -2) D) (-5, -2)

36. Which transformation would map a line to itself? A) A translation in any direction B) A reflection over the line itself C) A rotation around any point on the line D) A dilation from any point not on the line

37. If a quadrilateral has rotational symmetry of order 2, which of the following must be true? A) It is a rectangle B) It has at least two lines of symmetry C) Its diagonals bisect each other D) It has four equal sides

38. What is the relationship between two triangles if one can be mapped to the other using only rigid transformations? A) They are similar B) They are congruent C) They have the same area D) They have the same perimeter

39. A dilation with scale factor 3 followed by a dilation with scale factor 1/3, both from the same center, results in: A) A dilation with scale factor 1 B) A dilation with scale factor 0 C) A dilation with scale factor 3/1 D) A reflection

40. The point (4, -3) is reflected over the line y = x. What are the coordinates of the image? A) (-4, 3) B) (3, -4) C) (-3, 4) D) (-4, -3)

41. Which transformation changes the shape of a figure? A) Translation B) Reflection C) Rotation D) Dilation with different scale factors in the x and y directions

42. The composition of two reflections over intersecting lines results in: A) A translation B) A rotation C) Another reflection D) A dilation

43. If a figure has rotational symmetry of order 6, through what angle can it be rotated to map onto itself? A) 30° B) 45° C) 60° D) 90°

44. Under a transformation with rule $(x, y) \rightarrow (2x, 2y)$, the perimeter of a figure is: A) Doubled B) Quadrupled C) Increased by 2 D) Unchanged

45. Which of the following does not have an inverse transformation? A) Translation B) Reflection C) Rotation D) All have inverse transformations

46. The line of reflection that maps the point (3, 5) to (5, 3) is: A) The x-axis B) The y-axis C) The line y = x D) The line y = -x

47. What is the image of a line with slope 2 after a reflection over the x-axis? A) A line with slope 2 B) A line with slope -2 C) A line with slope 1/2 D) A line with slope -1/2

48. A transformation that preserves collinearity, distances, and angle measures is called: A) A rigid transformation B) An affine transformation C) A projective transformation D) A topological transformation

49. Which transformation maps the point (a, b) to (-a, -b)? A) Reflection over the x-axis B) Reflection over the y-axis C) Rotation of 180° around the origin D) Translation by $(-2a, -2b)$

50. If a transformation maps a triangle with vertices at (0, 0), (1, 0), and (0, 1) to a triangle with vertices at (0, 0), (3, 0), and (0, 3), what transformation was applied? A) Translation by (0, 0) B) Rotation of 90° around the origin C) Reflection over the line y = x D) Dilation from the origin with scale factor 3

Congruence and Similarity

1. Which of the following is NOT a valid triangle congruence criterion? A) SSS B) SAS C) SSA D) ASA

2. In an isosceles triangle, which of the following statements is true? A) All three sides are congruent B) All three angles are congruent C) At least two sides are congruent D) No sides are congruent

3. If triangle ABC is similar to triangle DEF with a similarity ratio of 2:3, and the perimeter of triangle ABC is 12 units, what is the perimeter of triangle DEF? A) 8 units B) 18 units C) 36 units D) 24 units

4. If two triangles are congruent, which of the following is NOT necessarily true? A) They have the same area B) They have the same perimeter C) They have the same orientation D) They have the same angle measures

5. In triangle ABC, the coordinates of the vertices are A(1, 2), B(4, 3), and C(2, 5). Which type of triangle is this? A) Equilateral B) Isosceles C) Scalene D) Right

6. If triangle ABC has sides of lengths 7, 24, and 25, what type of triangle is it? A) Right B) Obtuse C) Acute D) Equilateral

7. If triangles ABC and DEF are similar with a scale factor of 3, what is the ratio of the area of triangle ABC to the area of triangle DEF? A) 1:3 B) 3:1 C) 1:9 D) 9:1

122

8. Which of the following is a valid criterion for proving that two triangles are similar? A) AAA B) SSA C) AAS D) ASS

9. The altitudes of a triangle intersect at a point called the: A) Centroid B) Orthocenter C) Circumcenter D) Incenter

10. In the figure below, triangles ABC and DEF are shown. Which congruence criterion can be used to prove that these triangles are congruent? A) SSS B) SAS C) ASA D) AAS

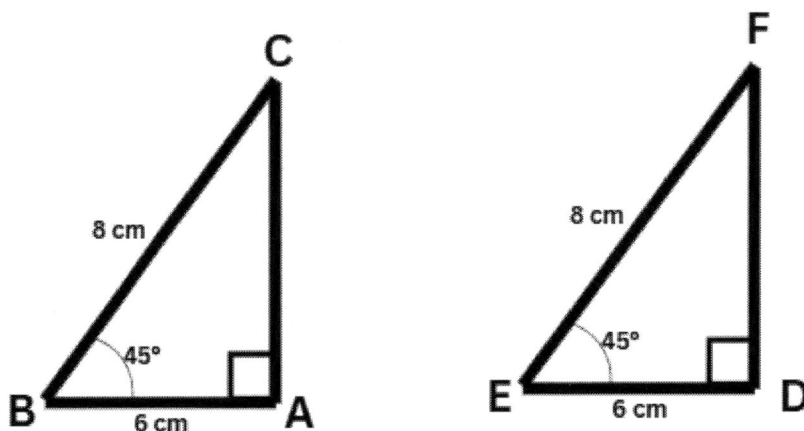

11. If two triangles are similar with a ratio of corresponding sides 2:5, what is the ratio of their areas? A) 2:5 B) 4:25 C) 4:10 D) 5:2

12. What is the measure of each interior angle in an equilateral triangle? A) 30° B) 45° C) 60° D) 90°

13. A midsegment of a triangle: A) Connects a vertex to the midpoint of the opposite side B) Connects the midpoints of two sides of the triangle C) Bisects the corresponding angle D) Is perpendicular to one of the sides

14. If triangle ABC ~ triangle DEF, which of the following is true? A) AB = DE B) AB/DE = BC/EF = AC/DF C) Area of triangle ABC = Area of triangle DEF D) Perimeter of ABC / Perimeter of DEF = 1

15. The perpendicular bisectors of the sides of a triangle intersect at a point called the: A) Centroid B) Orthocenter C) Circumcenter D) Incenter

16. In a right triangle, if one of the acute angles measures 30°, what is the measure of the other acute angle? A) 30° B) 45° C) 60° D) 90°

17. In triangle PQR, PR = QR and angle P = 52°. What is the measure of angle Q? A) 38° B) 52° C) 64° D) 76°

18. Which condition does NOT guarantee that a triangle is isosceles? A) Two sides are congruent B) Two angles are congruent C) The altitude to one side bisects that side D) One angle measures 60°

19. If triangle ABC is congruent to triangle DEF, and the area of triangle ABC is 24 square units, what is the area of triangle DEF? A) 8 square units B) 12 square units C) 24 square units D) 48 square units

20. What is the Triangle Inequality Theorem? A) The sum of the angles in a triangle is 180° B) The sum of the lengths of any two sides of a triangle is less than the length of the third side C) The sum of the lengths of any two sides of a triangle is greater than the length of the third side D) In a right triangle, the square of the hypotenuse equals the sum of the squares of the other two sides

21. If triangle ABC has angle measures of 30°, 60°, and 90°, what type of triangle is it? A) Equilateral B) Isosceles C) Scalene D) Obtuse

22. In an equilateral triangle, what special point coincides with the centroid, orthocenter, circumcenter, and incenter? A) The midpoint of one side B) One of the vertices C) The center of the triangle D) Such a point does not exist

23. In triangle ABC, the angle bisector from vertex A divides the opposite side BC into segments of lengths 4 and 5. If AB = 6, what is the length of AC? A) 7.5 B) 9 C) 10 D) 12

24. Which of the following transformations preserves the size and shape of a figure? A) Dilation B) Reflection C) Shear D) Stretching

25. If triangle ABC has sides of lengths 3, 4, and 5, what is its area? A) 6 square units B) 7.5 square units C) 10 square units D) 12 square units

26. In the figure below, triangles PQR and STU are shown. Which similarity criterion can be used to prove that these triangles are similar? A) AA B) SAS C) SSS D) Not enough information is provided

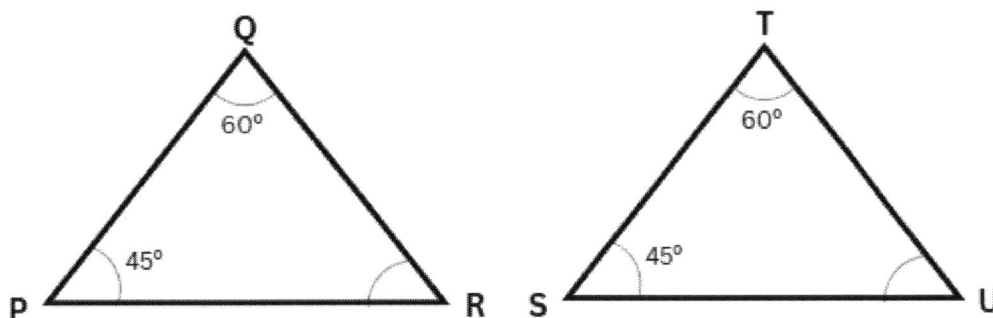

27. In a 45°-45°-90° triangle, if the length of the hypotenuse is 10 units, what is the length of each leg? A) 5 units B) $5\sqrt{2}$ units C) 10 units D) 7.07 units

28. Which of the following is true about the medians of a triangle? A) They are perpendicular to the sides of the triangle B) They bisect the angles of the triangle C) They intersect at the circumcenter D) They intersect at the centroid

29. If two triangles are congruent by the SAS criterion, which of the following must be congruent? A) Three pairs of sides B) Two pairs of sides and the included angle C) Two pairs of angles and the included side D) Three pairs of angles

30. The ratio of the areas of two similar triangles is 9:16. What is the ratio of their corresponding sides? A) 3:4 B) 9:16 C) 3:8 D) 9:4

31. The angle bisectors of a triangle intersect at a point called the: A) Centroid B) Orthocenter C) Circumcenter D) Incenter

32. In triangle ABC, if AB = 10, BC = 8, and AC = 6, which side is opposite to the largest angle? A) AB B) BC C) AC D) Cannot be determined

33. If two triangles are similar with a ratio of 2:3, and one triangle has an area of 12 square units, what is the area of the other triangle? A) 8 square units B) 18 square units C) 27 square units D) 16 square units

34. Which condition is sufficient to prove that two triangles are congruent? A) All three angles are congruent B) Two sides and a non-included angle are congruent C) Two angles and the non-included side are congruent D) Two sides are congruent

35. In a right triangle, the measure of one acute angle is twice the measure of the other acute angle. What are the angle measures of the triangle? A) 30°, 60°, 90° B) 45°, 45°, 90° C) 35°, 55°, 90° D) 20°, 70°, 90°

36. The centroid of a triangle divides each median in what ratio from the vertex to the midpoint of the opposite side? A) 1:1 B) 1:2 C) 2:1 D) 3:1

37. What is the minimum number of congruent parts needed to prove that two triangles are congruent? A) 2 B) 3 C) 4 D) 6

38. In a 30°-60°-90° triangle, if the shortest side is 5 units long, what is the length of the hypotenuse? A) $5\sqrt{2}$ units B) 10 units C) $5\sqrt{3}$ units D) 15 units

39. If a line divides two sides of a triangle proportionally, then it is: A) Parallel to the third side B) Perpendicular to the third side C) An angle bisector D) A median

40. The midsegment of a triangle connects the midpoints of two sides and is: A) Parallel to the third side and equal to half its length B) Perpendicular to the third side C) Equal in length to the third side D) Forming a 45° angle with both sides

41. In an isosceles triangle, the altitude to the base: A) Bisects the base B) Is equal in length to the base C) Forms a 45° angle with the base D) Is equal in length to one of the congruent sides

42. If a point is equidistant from the three vertices of a triangle, it is located at the: A) Centroid B) Orthocenter C) Circumcenter D) Incenter

43. If the perimeter of a triangle is 24 units and two sides measure 8 units and 10 units, what is the length of the third side? A) 6 units B) 8 units C) 10 units D) 12 units

44. The shortest distance from a point to a line is along: A) A line parallel to the given line B) The perpendicular from the point to the line C) An angle bisector from the point to the line D) Any line connecting the point to the given line

45. In a triangle with sides of lengths 8, 15, and 17, which of the following is true? A) The triangle is equilateral B) The triangle is isosceles C) The triangle is right D) The triangle is obtuse

46. If two angles of a triangle measure 55° and 65°, what is the measure of the third angle? A) 50° B) 60° C) 70° D) 80°

47. A triangle has sides of lengths 9, 9, and 12. What type of triangle is it? A) Equilateral B) Isosceles C) Scalene D) Right

48. In a triangle, if the three medians have lengths 9, 12, and 15, what is the perimeter of the triangle? A) 24 B) 30 C) 36 D) Cannot be determined from the given information

49. If triangle ABC is similar to triangle DEF with a similarity ratio of 3:4, and AC = 6, what is the length of DF? A) 4.5 B) 8 C) 24 D) Cannot be determined without more information

50. If a triangle has vertices at (0, 0), (3, 0), and (0, 4), what is its area? A) 3 square units B) 6 square units C) 12 square units D) 24 square units

---

Polygons and Quadrilaterals

---

1. What is the sum of the interior angles in a hexagon? A) 360° B) 540° C) 720° D) 1080°

2. A quadrilateral has diagonals that bisect each other. This quadrilateral must be a: A) Rectangle B) Rhombus C) Parallelogram D) Square

3. In a regular octagon, what is the measure of each interior angle? A) 108° B) 120° C) 135° D) 144°

4. What is the value of x in the figure below? A) 30° B) 42° C) 50° D) 70°

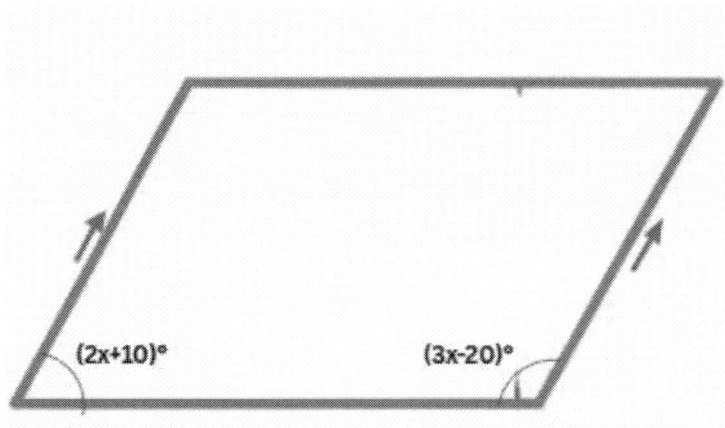

5. If one exterior angle of a regular polygon measures 24°, how many sides does the polygon have? A) 12 B) 15 C) 18 D) 24

6. The diagonals of which quadrilateral are always perpendicular to each other? A) Rectangle B) Parallelogram C) Rhombus D) Trapezoid

7. Which of the following is NOT a property of a parallelogram? A) Opposite sides are parallel B) Opposite angles are congruent C) Diagonals bisect each other D) Diagonals are perpendicular

8. In a quadrilateral, three angles measure 75°, 95°, and 110°. What is the measure of the fourth angle? A) 80° B) 85° C) 90° D) 100°

9. What is the area of a rhombus with diagonals measuring 8 cm and 12 cm? A) 24 cm² B) 48 cm² C) 96 cm² D) 192 cm²

10. In a regular pentagon, what is the measure of each exterior angle? A) 36° B) 54° C) 72° D) 108°

11. Which of the following statements about a square is FALSE? A) It is a parallelogram B) Its diagonals are congruent C) All sides are congruent D) Its diagonals form four congruent triangles

12. A quadrilateral has exactly one pair of parallel sides. This quadrilateral is a: A) Parallelogram B) Rhombus C) Trapezoid D) Square

13. What is the perimeter of a regular hexagon with a side length of 5 cm? A) 15 cm B) 25 cm C) 30 cm D) 60 cm

14. The sum of the exterior angles of any convex polygon is: A) 180° B) 360° C) Depends on the number of sides D) Depends on the interior angle sum

15. In a trapezoid with bases 8 cm and 14 cm, what is the length of the midsegment? A) 6 cm B) 11 cm C) 16 cm D) 22 cm

16. Which of the following is NOT a property of a rhombus? A) All sides are congruent B) Opposite sides are parallel C) All angles are congruent D) Diagonals bisect each other

127

17. If a quadrilateral has four congruent sides, it must be a: A) Square B) Rectangle C) Rhombus D) Parallelogram

18. What is the interior angle sum formula for an n-sided polygon? A) (n - 2) × 90° B) (n - 2) × 180° C) n × 180° D) (n + 2) × 180°

19. How many diagonals can be drawn in a convex polygon with 9 sides? A) 9 B) 18 C) 27 D) 36

20. In an isosceles trapezoid, which of the following is TRUE? A) All sides are congruent B) The diagonals are congruent C) All angles are congruent D) The bases are perpendicular to the legs

21. What is the area of a trapezoid with bases 5 cm and 11 cm and a height of 4 cm? A) 20 cm² B) 32 cm² C) 44 cm² D) 55 cm²

22. Which of the following is the correct formula for finding the number of diagonals in a polygon with n sides? A) $n(n - 1)/2$ B) $n(n - 3)/2$ C) $n(n - 2)$ D) $n^2 - n$

23. If a regular polygon has each interior angle measuring 140°, how many sides does it have? A) 9 B) 10 C) 12 D) 18

24. The diagonals of a kite: A) Are congruent B) Bisect each other C) Are perpendicular D) Form four congruent triangles

25. In a regular hexagon, what is the ratio of a side length to the radius of the circumscribed circle? A) 1:1 B) 1:$\sqrt{3}$ C) 1:2 D) 2:3

26. What is the area of a regular hexagon with side length 4 units? A) 24 square units B) 24√3 square units C) 48 square units D) 96 square units

27. A square with side length 6 cm is a special case of which of the following? A) Trapezoid only B) Rhombus only C) Rectangle only D) Both rhombus and rectangle

28. If a pentagon has interior angles measuring 100°, 110°, 120°, and 130°, what is the measure of the fifth angle? A) 80° B) 100° C) 160° D) 180°

29. Which of the following is NOT a property of a rectangle? A) Opposite sides are congruent B) Diagonals are congruent C) Diagonals are perpendicular D) All angles are right angles

30. What is the measure of each interior angle in a regular 20-sided polygon? A) 156° B) 162° C) 168° D) 174°

31. In a parallelogram with adjacent angles measuring 70° and 110°, what are the measures of the other two angles? A) 70° and 70° B) 70° and 110° C) 110° and 110° D) 80° and 100°

32. Which of the following properties can be used to prove that a quadrilateral is a parallelogram? A) One pair of opposite sides is congruent B) All sides are congruent C) Both pairs of opposite sides are parallel D) The diagonals are perpendicular

33. The polygon shown has how many lines of symmetry? A) 3 B) 4 C) 6 D) 12

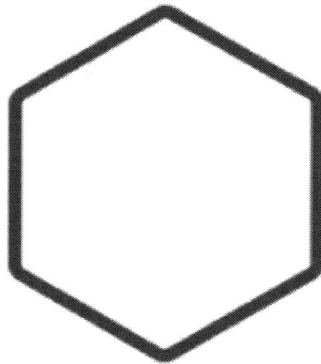

34. What is the perimeter of a rhombus with diagonals measuring 10 cm and 24 cm? A) 34 cm B) 40 cm C) 52 cm D) 68 cm

35. In which quadrilateral are the diagonals always perpendicular bisectors of each other? A) Rectangle B) Parallelogram C) Square D) Rhombus

36. What is the formula for the area of a regular polygon with n sides, side length s, and apothem a? A) $(1/2) \times n \times s \times a$ B) $n \times s \times a$ C) $n \times s^2$ D) $(n \times s^2)/4$

37. If a quadrilateral has exactly one pair of opposite sides that are both parallel and congruent, it is: A) Always a parallelogram B) Always a trapezoid C) Either a parallelogram or a trapezoid D) Neither a parallelogram nor a trapezoid

38. The apothem of a regular hexagon with side length 8 cm is: A) 4 cm B) $4\sqrt{3}cm$ C) 8 cm D) $8\sqrt{3}cm$

39. Which statement is TRUE about the diagonals of a trapezoid? A) They are always perpendicular B) They always bisect each other C) They are always congruent D) None of the above

40. A quadrilateral with four congruent sides and four right angles is: A) A rhombus B) A rectangle C) A square D) A parallelogram

41. In a kite, which of the following is TRUE? A) All sides are congruent B) Two pairs of adjacent sides are congruent C) Opposite sides are congruent D) All angles are congruent

42. What is the interior angle sum of a 13-sided polygon? A) 1980° B) 2160° C) 2340° D) 2520°

43. The midsegment of a trapezoid is: A) Parallel to the bases B) Perpendicular to the legs C) Equal to the sum of the bases D) Equal to the height of the trapezoid

44. In a rhombus with side length 5 cm and one angle measuring 60°, what is the area? A) 25 cm² B) $25\sqrt{3}$ cm² C) 50 cm² D) $50\sqrt{3}cm²$

45. If the diagonals of a quadrilateral bisect each other and are perpendicular, the quadrilateral must be a: A) Rectangle B) Rhombus C) Square D) Kite

46. What is the minimum number of sides a regular polygon can have? A) 3 B) 4 C) 5 D) 6

47. In a regular octagon, what is the measure of each exterior angle? A) 36° B) 45° C) 60° D) 135°

48. Which of the following cannot be constructed using a compass and straightedge? A) Regular triangle B) Regular square C) Regular pentagon D) Regular heptagon

49. The formula for the number of triangles formed by connecting all diagonals in a convex n-sided polygon is: A) $n - 2$ B) $n - 3$ C) n^2 D) $n(n-3)/2$

50. If the diagonals of a parallelogram are perpendicular, what type of quadrilateral must it be? A) Rectangle B) Rhombus C) Square D) Kite

Right Triangles and Trigonometry

1. In a right triangle, if one leg measures 8 units and the hypotenuse measures 17 units, what is the length of the other leg? A) 9 units B) 15 units C) 12 units D) 16 units

2. In a 30°-60°-90° triangle, if the shortest side measures 5 units, what is the length of the hypotenuse? A) $5\sqrt{2}$ units B) $5\sqrt{3}$ units C) 10 units D) $10\sqrt{3}$ units

3. If sin θ = 3/5, what is the value of cos θ? A) 4/5 B) 3/4 C) 3/5 D) 5/3

4. A ladder 20 feet long leans against a vertical wall. If the bottom of the ladder is 12 feet from the base of the wall, how high up the wall does the ladder reach? A) 8 feet B) 16 feet C) 10 feet D) 18 feet

5. In a right triangle, the legs measure 5 units and 12 units. What is the area of the triangle? A) 30 square units B) 60 square units C) 25 square units D) 17 square units

6. If tan θ = 4/3, what is the value of sin θ? A) 4/5 B) 3/5 C) 3/4 D) 5/4

7. In a 45°-45°-90° triangle, if each leg measures 6 units, what is the length of the hypotenuse? A) 6 units B) $6\sqrt{2}$ units C) 12 units D) $3\sqrt{2}$ units

8. In the right triangle shown below, what is the value of x? A) 4 B) 5 C) 6 D) 8

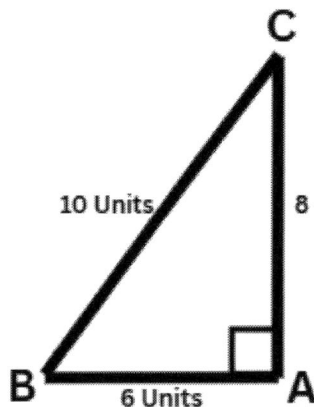

9. If cos θ = 5/13, what is the value of tan θ? A) 5/12 B) 12/5 C) 13/5 D) 12/13

10. In triangle ABC, if a = 7, b = 24, and c = 25, what type of triangle is it? A) Acute B) Obtuse C) Right D) Equilateral

11. Which of the following is not a Pythagorean triple? A) (3, 4, 5) B) (5, 12, 13) C) (6, 8, 10) D) (7, 8, 11)

12. The angle of elevation from a point on the ground to the top of a building is 32°. If the point is 50 meters from the base of the building, how tall is the building to the nearest meter? A) 31 meters B) 26 meters C) 42 meters D) 16 meters

13. In a right triangle, if the sine of one acute angle is 3/5, what is the cosine of the other acute angle? A) 3/5 B) 4/5 C) 5/3 D) 4/3

14. If sin θ = 12/13, what is the value of cos θ? A) 5/13 B) 12/13 C) 5/12 D) 12/5

15. In a 30°-60°-90° triangle, if the hypotenuse is 16 units, what is the length of the side opposite to the 30° angle? A) 16 units B) 8 units C) $8\sqrt{3}$ units D) $16\sqrt{3}$ units

16. Using the Law of Sines, in triangle ABC, if sin A = sin B and side a = 7, what is the length of side b? A) Cannot be determined B) 7 C) 14 D) 3.5

17. What is the value of tan 45°? A) 0 B) 1 C) $\sqrt{3}$ D) $1/\sqrt{3}$

18. In a right triangle, if one angle is 25°, what is the measure of the other acute angle? A) 25° B) 65° C) 115° D) 155°

19. If leg a = 8 and leg b = 15 in a right triangle, what is the measure of the angle opposite to leg a? A) $\sin^{-1}(8/17)$ B) $\cos^{-1}(8/17)$ C) $\tan^{-1}(8/15)$ D) $\cos^{-1}(15/17)$

20. Using the Law of Cosines, in triangle ABC with sides a = 6, b = 8, and c = 10, what is the measure of angle C to the nearest degree? A) 60° B) 90° C) 120° D) 30°

131

21. A guy wire is attached to the top of a 50-foot tall pole and is anchored 30 feet from the base of the pole. What is the length of the guy wire? A) 80 feet B) 58.3 feet C) 60 feet D) 40 feet

22. Which of the following values of x would make 3, 4, x the side lengths of a right triangle? A) 5 B) 6 C) 7 D) 8

23. In a right triangle, if cos θ = 7/25, what is the value of tan θ? A) 24/7 B) 7/24 C) 25/24 D) 24/25

24. If the legs of a right triangle measure 5 units and x units, and the hypotenuse measures (x+2) units, what is the value of x? A) 3 B) 4 C) 5 D) 6

25. In the diagram below, what is the measure of angle θ? A) 30° B) 45° C) 60° D) 75°

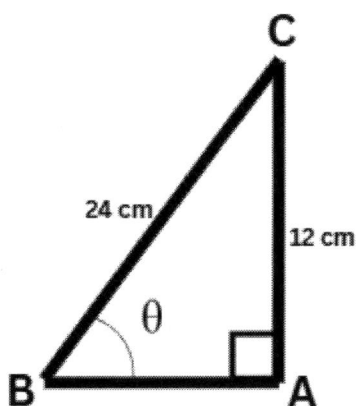

26. The value of sin 30° is: A) 1/4 B) 1/2 C) √3/2 D) 1

27. In a right triangle, if tan θ = 3/4, what is the measure of θ to the nearest degree? A) 37° B) 53° C) 36° D) 43°

28. What is the value of cos 60°? A) 1/2 B) √3/2 C) 1/√3 D) √3

29. If the hypotenuse of a right triangle is 10 units and one leg is 8 units, what is the length of the other leg? A) 2 units B) 4 units C) 6 units D) 8 units

30. Using the Law of Cosines, what is the measure of the largest angle in a triangle with sides 7, 8, and 13? A) About 30° B) About 60° C) About 90° D) About 120°

31. What is the sine of an angle in a right triangle if the opposite side is 15 and the hypotenuse is 25? A) 3/5 B) 4/5 C) 15/25 D) 25/15

32. The value of cos 45° is: A) 1 B) 1/2 C) √2/2 D) √3/2

33. In a triangle, if the Law of Sines gives two possible solutions, this situation is known as: A) The congruent case B) The ambiguous case C) The complementary case D) The triangle inequality case

34. A right triangle has one acute angle measuring 40°. What is the measure of the other acute angle? A) 40° B) 50° C) 140° D) 130°

35. In a 45°-45°-90° triangle, the ratio of a leg to the hypotenuse is: A) 1:$\sqrt{2}$ B) 1:2 C) 1:$\sqrt{3}$ D) 1:1

36. If the sides of a triangle measure 7, 9, and 11 units, what type of triangle is it? A) Right B) Obtuse C) Acute D) Isosceles

37. The geometric mean of two positive numbers a and b is: A) (a + b)/2 B) $\sqrt{(ab)}$ C) ab D) a/b

38. If sin A = 4/5 and sin B = 5/13, what is sin(A + B)? A) 9/13 B) 45/65 C) 63/65 D) 77/65

39. Which of the following cannot be the measures of the three angles in a triangle? A) 30°, 60°, 90° B) 45°, 45°, 90° C) 30°, 60°, 100° D) 30°, 30°, 120°

40. The tangent of an angle in a right triangle equals: A) Adjacent side ÷ Opposite side B) Opposite side ÷ Adjacent side C) Opposite side ÷ Hypotenuse D) Adjacent side ÷ Hypotenuse

41. What is the area of a triangle with sides a = 13, b = 14, and c = 15? A) 84 square units B) 91 square units C) 126 square units D) 182 square units

42. If the legs of a right triangle have lengths a and b, and the hypotenuse has length c, which of the following relations is always true? A) a + b > c B) a + b = c C) a + b < c D) $a^2 + b^2 = c^2$

43. In a right triangle, if one leg measures x and the other leg measures 2x, the hypotenuse measures: A) 3x B) $2\sqrt{2}x$ C) $x\sqrt{5}$ D) 3x/2

44. In a right triangle, the altitude to the hypotenuse creates two similar triangles that are: A) Similar to each other only B) Similar to the original triangle only C) Similar to each other and to the original triangle D) Not similar to any other triangle

45. The solution(s) to the equation sin θ = 1.5 is/are: A) θ = 90° B) θ = 30° C) No solution D) θ = 180°

46. Using the Law of Sines, in triangle ABC, if A = 30°, B = 45°, and a = 10, what is the length of side b? A) $10\sqrt{2}$ B) 14.1 C) 7.1 D) 10

47. The area of a triangle can be calculated using: A) A = ab sin C B) A = 1/2 × ab × sin C C) A = a + b + c D) A = ab cos C

48. If a 15-foot ladder leaning against a wall makes a 60° angle with the ground, how high up the wall does the ladder reach, to the nearest foot? A) 13 feet B) 7 feet C) 8 feet D) 15 feet

49. In a triangle with sides a, b, and c, the Law of Cosines states that: A) $a^2 = b^2 + c^2 - 2bc \cos A$ B) $a^2 = b^2 + c^2 + 2bc \cos A$ C) $a^2 = b^2 - c^2 + 2bc \cos A$ D) a = b + c - 2bc cos A

50. A triangle has sides of lengths 7, 8, and 9 units. The measure of the angle opposite the side of length 9 is approximately: A) 45° B) 60° C) 75° D) 90°

Circles

1. Which of the following is the equation of a circle with center (3, -2) and radius 5? A) $(x - 3)^2 + (y + 2)^2 = 5$ B) $(x - 3)^2 + (y + 2)^2 = 25$ C) $(x + 3)^2 + (y - 2)^2 = 25$ D) $(x + 3)^2 + (y - 2)^2 = 5$

2. A circle has a radius of 6 cm. What is its circumference? A) 12π cm B) 36π cm C) 6π cm D) 24π cm

3. In a circle, a central angle of 72° corresponds to an arc length of 8π cm. What is the radius of the circle? A) 20 cm B) 72 cm C) 40 cm D) 36 cm

4. A sector of a circle has a central angle of 60° and radius of 12 cm. What is the area of the sector? A) 24π cm² B) $144\pi/3$ cm² C) 48π cm² D) 24 cm²

5. A circle has the equation $x^2 + y^2 + 4x - 6y + 9 = 0$. What is its center? A) (2, 3) B) (-2, 3) C) (2, -3) D) (-2, -3)

6. What is the length of a chord that is 8 cm from the center of a circle with radius 10 cm? A) 6 cm B) 8 cm C) 12 cm D) 16 cm

7. If a circle has a diameter of 14 cm, what is its area? A) 49π cm² B) 14π cm² C) 196π cm² D) 28π cm²

8. In the figure below, O is the center of the circle, and angle AOB = 130°. What is the measure of arc AB? A) 65° B) 130° C) 50° D) 230°

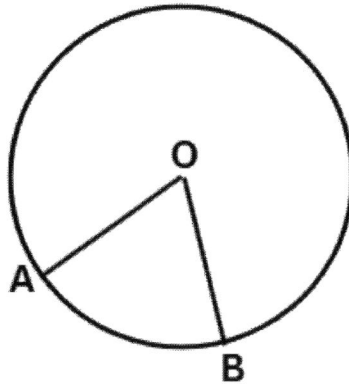

9. A tangent line to a circle is: A) Perpendicular to the radius at the point of tangency B) Parallel to the radius at the point of tangency C) Always vertical to the x-axis D) Always horizontal to the y-axis

10. The equation of a circle with center at the origin and passing through the point (0, -7) is: A) $x^2 + y^2 = 7$ B) $x^2 + y^2 = 49$ C) $x^2 + y^2 = -7$ D) $x^2 + y^2 = -49$

11. If two chords of a circle are congruent, then they are: A) Equidistant from the center B) Perpendicular to each other C) Equal in measure to the diameter D) Always diameters

12. If a circle has center (4, -1) and passes through the point (7, 3), what is its radius? A) 5 B) 25 C) $\sqrt{25}$ D) $3\sqrt{5}$

13. If a circle has the equation $(x - 3)^2 + (y + 5)^2 = 36$, what is its radius? A) 6 B) 36 C) $\sqrt{36}$ D) 6^2

14. What is the relationship between an inscribed angle and its intercepted arc in a circle? A) The inscribed angle equals the measure of its intercepted arc B) The inscribed angle equals half the measure of its intercepted arc C) The inscribed angle equals twice the measure of its intercepted arc D) The inscribed angle equals 90° minus the measure of its intercepted arc

15. The standard form of the equation of a circle is: A) $Ax^2 + By^2 + Cx + Dy + E = 0$ B) $(x - h)^2 + (y - k)^2 = r^2$ C) $(h - x)^2 + (k - y)^2 = r^2$ D) $x^2 + y^2 = r^2$

16. If a central angle of 1 radian intercepts an arc of length 12 cm, what is the radius of the circle? A) 12 cm B) 6 cm C) 24 cm D) 12π cm

17. In a circle with center O, if chord AB is not a diameter, which of the following statements is always true? A) Chord AB passes through O B) The perpendicular bisector of chord AB passes through O C) O is the midpoint of chord AB D) Chord AB is tangent to the circle

18. What is the area of a segment of a circle with radius 10 cm and central angle 90°? A) 25π cm² - 50 cm² B) 25π cm² C) 50π cm² D) 100 cm²

19. If two circles are tangent externally, what is the relationship between the distance between their centers and their radii? A) The distance equals the sum of the radii B) The distance equals the product of the radii C) The distance equals the difference of the radii D) The distance equals twice the sum of the radii

20. A secant line to a circle: A) Intersects the circle at exactly one point B) Intersects the circle at exactly two points C) Is tangent to the circle D) Does not intersect the circle

21. What is the equation of the circle with center (0, 0) and diameter 16? A) $x^2 + y^2 = 16$ B) $x^2 + y^2 = 8$ C) $x^2 + y^2 = 64$ D) $x^2 + y^2 = 4$

22. If two circles intersect at two points, what is the relationship between the line connecting their centers and the line containing the points of intersection? A) They are parallel B) They are perpendicular C) They form a 45° angle D) They are the same line

23. In a circle, if two chords intersect, which of the following statements is true about the segments of the chords? A) The sum of the products of the segments of each chord is equal B) The difference of the products of the segments of each chord is equal C) The product of one chord and a segment of the other chord is constant D) The segments of each chord are equal in length

135

24. If a tangent and a secant are drawn to a circle from an external point, what is the relationship between the tangent segment and the segments of the secant? A) The tangent segment equals the product of the secant segments B) The square of the tangent segment equals the product of the entire secant and its external part C) The tangent segment equals the sum of the secant segments D) The square of the tangent segment equals the sum of the squares of the secant segments

25. A chord of length 12 cm is drawn in a circle of radius 13 cm. What is the distance from the center of the circle to the chord? A) 5 cm B) 7 cm C) 12 cm D) 13 cm

26. In a Circle, point P is outside the circle, and PA and PB are tangent to the circle. If PA = 8 cm, what is PB? A) 4 cm B) 8 cm C) 16 cm D) It cannot be determined from the given information

27. What is the area of a circle with a circumference 18π cm? A) 81π cm² B) 9π cm² C) 81 cm² D) 18π cm²

28. If a circle has equation $x^2 + y^2 - 4x + 6y - 12 = 0$, what is its radius? A) 5 B) 25 C) $\sqrt{25}$ D) 12

29. What is the measure of an inscribed angle that intercepts a semicircle? A) 90° B) 180° C) 270° D) 360°

30. Two concentric circles have radii of 3 cm and 7 cm. What is the area of the ring between them? A) 4π cm² B) 10π cm² C) 16π cm² D) 40π cm²

31. If two circles have radii 5 cm and 12 cm, and their centers are 13 cm apart, how are these circles positioned relative to each other? A) One circle is completely inside the other B) The circles intersect at two points C) The circles are tangent internally D) The circles are tangent externally

32. If a circle has the equation $x^2 + y^2 - 6x + 8y + 15 = 0$, what are the coordinates of its center? A) (3, -4) B) (-3, 4) C) (3, 4) D) (-3, -4)

33. Which of the following is not equivalent to the others? A) The tangent to a circle is perpendicular to the radius at the point of tangency B) If a line is perpendicular to a radius at its endpoint on the circle, then the line is tangent to the circle C) If a tangent and a radius of a circle are perpendicular, then they must intersect at a point on the circle D) If a line intersects a circle at exactly one point, it must be perpendicular to the radius at that point

34. What is the length of the tangent segment from a point 15 cm away from the center of a circle with radius 9 cm? A) 6 cm B) 12 cm C) 18 cm D) 24 cm

35. The area of a sector with central angle $3\pi/4$ radians in a circle of radius 6 cm is: A) 9π cm² B) 18π cm² C) $27\pi/2$ cm² D) 36π cm²

36. If two circles have the same center, they are called: A) Tangent circles B) Concentric circles C) Congruent circles D) Similar circles

37. If the center of a circle is at (-2, 3) and its radius is 5, which of the following points lies on the circle? A) (2, 3) B) (-2, 8) C) (1, 7) D) (3, 3)

38. The general form of the equation of a circle is: A) $(x - h)^2 + (y - k)^2 = r^2$ B) $Ax^2 + By^2 + Cx + Dy + E = 0$ where $A = B \neq 0$ C) $Ax^2 + By^2 + Cx + Dy + E = 0$ where $A = B = 1$ D) $x^2 + y^2 + Dx + Ey + F = 0$

39. If a circle has its center at the origin, which of these points could not lie on the circle? A) (3, 4) B) (5, 0) C) (0, -5) D) (4, 4)

40. What is the equation of a circle with center (2, -3) and passing through the origin? A) $(x - 2)^2 + (y + 3)^2 = 13$ B) C) $(x + 2)^2 + (y - 3)^2 = 13$ D) $(x - 2)^2 + (y + 3)^2 = 4$

41. If angle ABC is an inscribed angle in a circle and intercepts an arc of 140°, what is the measure of angle ABC? A) 140° B) 70° C) 280° D) 35°

42. In a circle with center O, if chord AB subtends a central angle of 60° at O, and the radius of the circle is 8 cm, what is the length of chord AB? A) 4 cm B) 8 cm C) 8√3 cm D) 16 cm

43. If a secant and a tangent to a circle are drawn from an external point, what is the measure of the angle between them? A) Equal to the inscribed angle that intercepts the same arc B) Half the measure of the intercepted arc C) Twice the measure of the intercepted arc D) Equal to the central angle that intercepts the same arc

44. What is the equation of the circle with diameter from (2, 5) to (-4, 1)? A) $(x - 1)^2 + (y - 3)^2 = 18$ B) $(x + 1)^2 + (y - 3)^2 = 18$ C) $(x - 1)^2 + (y - 3)^2 = \sqrt{18}$ D) $(x - 1)^2 + (y - 3)^2 = 9$

45. The circumference of a circle with radius r is: A) πr B) $2\pi r$ C) πr^2 D) $2\pi r^2$

46. A circle has radius 10 cm. What is the length of a 45° arc of this circle? A) $45\pi/18$ cm B) $5\pi/9$ cm C) 5π cm D) $10\pi/4$ cm

47. The perpendicular from the center of a circle to a chord: A) Bisects the chord B) Is equal in length to the radius C) Is equal in length to the chord D) Is always a diameter

48. If the equation of a circle is $x^2 + y^2 - 4x + 6y + 4 = 0$, what is the distance from the center of the circle to the origin? A) 5 B) √5 C) 5√2 D) 25

49. If two circles intersect, the maximum number of common tangent lines they can have is: A) 1 B) 2 C) 3 D) 4

50. A chord of length 10 cm is located 12 cm from the center of a circle. What is the radius of the circle? A) 13 cm B) 16 cm C) 17 cm D) 22 cm

Coordinate Geometry

1. What is the distance between points P(3, -4) and Q(-2, 5)? A) 5 B) 7 C) 10 D) √106

2. The midpoint of a line segment is (5, -2). If one endpoint is (8, 1), what are the coordinates of the other endpoint? A) (2, -5) B) (2, 3) C) (3, -5) D) (13, -3)

3. Which of the following is the equation of a line with slope 3 passing through the point (2, -4)? A) y = 3x - 10 B) y = 3x - 6 C) y = 3x + 2 D) y - 3x = -10

4. Two lines have slopes of -2 and 1/2, respectively. What is the relationship between these lines? A) They are parallel B) They are perpendicular C) They are the same line D) Cannot be determined from the given information

5. What is the slope of the line containing the points (5, 3) and (5, 9)? A) 0 B) 1 C) Undefined D) 6

6. Which of the following is the slope-intercept form of the line 3x - 2y = 12? A) y = 3/2x - 6 B) y = -3/2x + 6 C) y = 3/2x + 6 D) y = -3/2x - 6

7. What is the area of triangle ABC with vertices A(2, 1), B(6, 1), and C(4, 5)? A) 8 square units B) 12 square units C) 16 square units D) 20 square units

8. What is the equation of a circle with center (3, -2) and radius 5? A) $(x - 3)^2 + (y + 2)^2 = 5$ B) $(x - 3)^2 + (y + 2)^2 = 25$ C) $(x + 3)^2 + (y - 2)^2 = 25$ D) $(x - 3)^2 + (y - 2)^2 = 25$

9. The point (4, -3) lies in which quadrant of the coordinate plane? A) I B) II C) III D) IV

10. Which of the following is the point-slope form of a line passing through the point (-2, 5) with slope -3? A) y - 5 = -3(x + 2) B) y - 5 = -3(x - 2) C) y + 5 = -3(x + 2) D) y + 5 = -3(x - 2)

11. Two lines are perpendicular if the product of their slopes equals: A) 0 B) 1 C) -1 D) infinity

12. What is the distance from the point (2, 3) to the line 3x + 4y = 25? A) 0 B) 2 C) 3 D) 5

13. The coordinates of the midpoint of a line segment are (3, -4). If one endpoint has coordinates (7, 2), what are the coordinates of the other endpoint? A) (-1, -10) B) (10, -2) C) (-1, -6) D) (5, -8)

14. Which of the following is the equation of a line perpendicular to y = 2x + 3 and passing through the point (4, -1)? A) y = -1/2x + 1 B) y = -1/2x + 3 C) y = -1/2x - 1 D) y = -1/2x + 2

15. If the endpoints of a diameter of a circle are (1, 2) and (7, 8), what are the coordinates of the center of the circle? A) (3, 5) B) (4, 5) C) (8, 10) D) (6, 6)

16. What is the distance from the point (3, 4) to the origin? A) 5 B) 7 C) $\sqrt{25}$ D) $3\sqrt{2}$

17. A line has the equation 2x - 5y = 10. What is its y-intercept? A) (0, -2) B) (0, 2) C) (5, 0) D) (-5, 0)

18. Which quadrilateral is formed by the points A(0, 0), B(6, 0), C(6, 4), and D(0, 4)? A) Square B) Rectangle C) Rhombus D) Parallelogram

19. What is the slope of a line perpendicular to the line with equation 4x + 3y = 12? A) -4/3 B) 3/4 C) 4/3 D) -3/4

20. In a coordinate line, if points A, B, and C lie on the same line and the distance from A to C is 12 units, what is the distance from A to B? A) 3 units B) 4 units C) 8 units D) 9 units

21. The line with equation $y = 3x - 7$ has: A) Slope 3 and y-intercept (0, -7) B) Slope -7 and y-intercept (0, 3) C) Slope 3 and y-intercept (0, 7) D) Slope -3 and y-intercept (0, -7)

22. What is the standard form of the equation of a circle with center at (-1, 2) and passing through the point (3, -6)? A) $(x + 1)^2 + (y - 2)^2 = 100$ B) $(x - 1)^2 + (y + 2)^2 = 80$ C) $(x + 1)^2 + (y - 2)^2 = 80$ D) $(x - 1)^2 + (y - 2)^2 = 100$

23. If the coordinates of the midpoint of the line segment from A(p, q) to B(r, s) are (4, -3), which of the following must be true? A) $p + r = 8$ B) $p + r = 4$ C) $q + s = -6$ D) $q + s = 3$

24. What are the coordinates of the point that divides the line segment from P(2, -3) to Q(8, 5) in the ratio 2:1 (starting from P)? A) (4, -1) B) (5, 1) C) (6, 2) D) (4, 1)

25. Which of the following is true about the lines $2x + 3y = 6$ and $4x + 6y = 15$? A) They are parallel B) They are perpendicular C) They are the same line D) They intersect at a non-right angle

26. What is the center of the circle with the equation $x^2 + y^2 - 6x + 8y + 9 = 0$? A) (3, -4) B) (6, -8) C) (-3, 4) D) (-6, 8)

27. Which of the following represents a line parallel to $3x - 2y = 7$? A) $6x - 4y = 10$ B) $3x + 2y = 5$ C) $2x - 3y = 8$ D) $-3x + 2y = 9$

28. If points A(1, 3), B(5, 7), and C(3, -1) form a triangle, what is its area? A) 10 square units B) 12 square units C) 16 square units D) 20 square units

29. Which of the following is the equation of the perpendicular bisector of the line segment joining points (2, 5) and (6, 9)? A) $y = x + 3$ B) $y = -x + 13$ C) $y = x - 3$ D) $y = -x + 11$

30. What is the radius of the circle with equation $(x - 4)^2 + (y + 3)^2 = 49$? A) 7 B) 9 C) 16 D) 49

31. What is the distance between the parallel lines $2x - 3y = 6$ and $2x - 3y = -12$? A) 6 B) 3 C) $9/\sqrt{13}$ D) $18/\sqrt{13}$

32. The points (3, -1), (k, 5), and (7, 9) lie on the same line. What is the value of k? A) 4 B) 5 C) 6 D) 8

33. What is the standard form equation of the ellipse with center at (2, -3), a major axis of length 10 along the x-axis, and a minor axis of length 6? A) $(x - 2)^2/25 + (y + 3)^2/9 = 1$ B) $(x - 2)^2/9 + (y + 3)^2/25 = 1$ C) $(x + 2)^2/25 + (y - 3)^2/9 = 1$ D) $(x - 2)^2/100 + (y + 3)^2/36 = 1$

34. What is the equation of a line with x-intercept (3, 0) and y-intercept (0, -6)? A) $2x - y = 6$ B) $2x + y = 6$ C) $x - 2y = 3$ D) $x + 2y = 3$

35. If the distance between points P(3, k) and Q(7, 10) is 10, what is the value of k? A) 2 or 14 B) -2 or 18 C) 4 or 12 D) 6 or 14

139

36. The diagonals of a parallelogram intersect at the point (3, 2). If three of the vertices are (1, 5), (5, 5), and (1, -1), what are the coordinates of the fourth vertex? A) (5, -1) B) (5, -2) C) (6, -1) D) (7, -1)

37. Which of the following is true about the locus of points equidistant from the points (3, 5) and (7, 1)? A) It is a circle with a center at (5, 3) B) It is a line with slope 1 C) It is a line with slope -1 D) It is a parabola

38. If a parabola has its vertex at (2, -3) and its axis of symmetry is parallel to the y-axis, which of the following could be its equation? A) $y = (x-2)^2 - 3$ B) $x = (y+3)^2 + 2$ C) $y = -(x-2)^2 - 3$ D) $y = 2(x-2)^2 - 3$

39. What is the center of the hyperbola with equation $(x-4)^2/9 - (y+2)^2/16 = 1$? A) (4, -2) B) (4, 2) C) (-4, 2) D) (9, 16)

40. The slopes of two sides of a triangle are 2 and -3. What could be the slope of the third side? A) 6 B) 2/3 C) -2/3 D) Any of these could be correct

41. What is the distance between the point (-1, 3) and the line $y = 2x + 5$? A) 1 B) 2 C) 3 D) 4

42. What are the coordinates of the focus of the parabola with equation $y = -2(x-3)^2 + 4$? A) (3, 4.125) B) (3, 3.875) C) (3.125, 4) D) (2.875, 4)

43. Which of the following best describes the graph of the equation $|x - 3| + |y + 2| = 5$? A) Circle B) Square C) Rectangle D) Rhombus

44. If the graph of $2x^2 + 3y^2 = 12$ is an ellipse, what are the lengths of its semi-major and semi-minor axes? A) $2\sqrt{3}$ and $\sqrt{6}$ B) $\sqrt{6}$ and 2 C) 3 and 2 D) 6 and 4

45. What is the equation of the tangent line to the circle $x^2 + y^2 = 25$ at the point (3, 4)? A) $3x + 4y = 25$ B) $4x + 3y = 25$ C) $3x - 4y = 25$ D) $4x - 3y = 25$

46. If the vertices of a quadrilateral are at (1, 1), (4, 2), (5, 6), and (2, 5), what is its area? A) 14 square units B) 15 square units C) 16 square units D) 17 square units

47. What are the coordinates of the point that is the reflection of (-2, 5) across the line $y = x$? A) (5, -2) B) (-5, 2) C) (2, -5) D) (-5, -2)

48. Two circles have centers at (1, 2) and (4, 6) with radii 3 and 5, respectively. Which of the following describes their relationship? A) The circles are internally tangent B) The circles are externally tangent C) The circles intersect at two points D) One circle is inside the other without touching

49. What is the eccentricity of an ellipse with semi-major axis 5 and semi-minor axis 3? A) 4/5 B) 3/5 C) 2/5 D) 4/3

50. The equation $x^2 + y^2 + 6x - 8y + 25 = 0$ represents: A) A circle with radius 5 B) A circle with radius 0 C) An ellipse D) A point

Geometric Measurement and Dimension

1. What is the perimeter of a rectangle with length 12 cm and width 7 cm? A) 19 cm B) 38 cm C) 84 cm D) 42 cm

2. The circumference of a circle is 24π cm. What is the radius of the circle? A) 24 cm B) 12 cm C) 48 cm D) 6 cm

3. A regular hexagon has sides of length 5 units. What is its perimeter? A) 15 units B) 25 units C) 30 units D) 35 units

4. The area of a triangle is 36 square units. If the base is 12 units, what is the height? A) 3 units B) 6 units C) 9 units D) 12 units

5. What is the area of a circle with diameter 10 meters? A) 25π m² B) 100π m² C) 10π m² D) 5π m²

6. A trapezoid has parallel sides of lengths 8 cm and 14 cm, and a height of 6 cm. What is its area? A) 48 cm² B) 66 cm² C) 84 cm² D) 132 cm²

7. The area of a parallelogram is 120 square units. If its height is 8 units, what is the length of its base? A) 12 units B) 15 units C) 18 units D) 20 units

8. A sector of a circle has a central angle of 60° and radius of 12 cm. What is the area of the sector? A) 24π cm² B) 12π cm² C) 6π cm² D) 72π cm²

9. What is the volume of a cube with side length 5 inches? A) 15 in³ B) 25 in³ C) 75 in³ D) 125 in³

10. The volume of a rectangular prism is 360 cubic units. If its length is 10 units and its width is 6 units, what is its height? A) 3 units B) 6 units C) 9 units D) 12 units

11. In a triangular prism , the base is a right triangle with legs measuring 3 cm and 4 cm. If the height of the prism is 10 cm, what is its volume? A) 60 cm³ B) 120 cm³ C) 40 cm³ D) 30 cm³

12. A cylinder has radius 5 m and height 8 m. What is its volume? A) 40π m³ B) 200π m³ C) 100π m³ D) 400π m³

13. What is the surface area of a cube with edge length 4 units? A) 16 units² B) 64 units² C) 96 units² D) 128 units²

14. A cone has radius 6 cm and height 8 cm. What is its volume? A) 48π cm³ B) 96π cm³ C) 144π cm³ D) 288π cm³

15. The surface area of a sphere is 256π square units. What is its radius? A) 4 units B) 8 units C) 16 units D) 32 units

16. A rectangular pyramid has a base with length 8 cm and width 6 cm. If the height of the pyramid is 10 cm, what is its volume? A) 80 cm³ B) 160 cm³ C) 240 cm³ D) 480 cm³

17. What is the lateral surface area (excluding bases) of a cylinder with radius 3 m and height 7 m? A) 21π m² B) 42π m² C) 60π m² D) 30π m²

18. A sphere has radius 5 cm. What is its volume? A) 20π cm³ B) 125π cm³ C) $500\pi/3$ cm³ D) 100π cm³

19. The volume of a cone is 24π cubic units. If the radius of the base is 3 units, what is the height? A) 8 units B) 9 units C) 12 units D) 24 units

20. A square has a perimeter of 36 units. What is its area? A) 9 units² B) 18 units² C) 36 units² D) 81 units²

21. What is the area of a rhombus with diagonals of lengths 10 cm and 16 cm? A) 40 cm² B) 80 cm² C) 160 cm² D) 320 cm²

22. The volume of a rectangular prism increases by what factor when all of its dimensions are doubled? A) 2 B) 4 C) 6 D) 8

23. A regular pentagon has a perimeter of 45 units. What is the length of each side? A) 5 units B) 9 units C) 10 units D) 15 units

24. What is the surface area of a rectangular prism with length 5 cm, width 3 cm, and height 4 cm? A) 60 cm² B) 94 cm² C) 120 cm² D) $60\sqrt{2}$ cm²

25. If the radius of a sphere is doubled, by what factor does its volume increase? A) 2 B) 4 C) 6 D) 8

26. The surface area of a cylinder with radius 4 inches and height h inches is 96π square inches. What is the value of h? A) 6 inches B) 8 inches C) 10 inches D) 12 inches

27. A regular octagon has sides of length 4 units. What is its perimeter? A) 16 units B) 24 units C) 32 units D) 36 units

28. The lengths of the three sides of a triangle are 5 cm, 12 cm, and 13 cm. What is the area of the triangle? A) 30 cm² B) 32.5 cm² C) 37.5 cm² D) 40 cm²

29. What is the volume of a sphere with diameter 6 units? A) 36π units³ B) 72π units³ C) 113.04 units³ D) $36\pi/3$ units³

30. The base of a prism is a triangle with sides of lengths 5 cm, 12 cm, and 13 cm. If the height of the prism is 9 cm, what is its volume? A) 30 cm³ B) 270 cm³ C) 540 cm³ D) 585 cm³

31. A square has an area of 121 square units. What is its perimeter? A) 11 units B) 22 units C) 44 units D) 121 units

32. In the composite figure shown below (a rectangle with a semicircle on one end), the rectangle measures 8 cm by 4 cm. What is the perimeter of the entire figure? A) 24 cm B) 28 cm C) 28 + 4π cm D) 24 + 4π cm

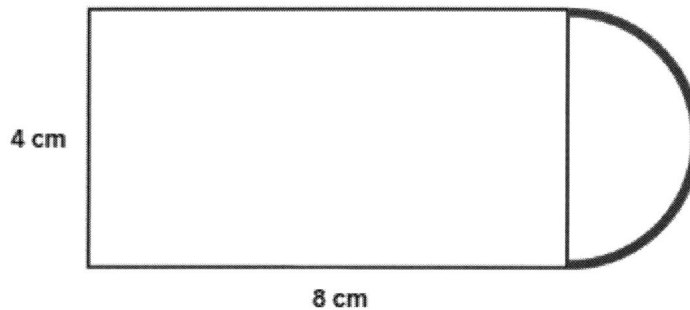

4 cm

8 cm

33. What is the lateral surface area (excluding bases) of a regular square pyramid with side length 6 units and slant height 5 units? A) 30 units² B) 60 units² C) 120 units² D) 180 units²

34. If a cube has a surface area of 96 square units, what is its volume? A) 16 units³ B) 32 units³ C) 64 units³ D) 512 units³

35. The volume of a cone is 40π cubic cm. If the height is 15 cm, what is the radius of the base? A) 2 cm B) 3 cm C) 4 cm D) 8 cm

36. A cylinder and a cone have the same base radius and the same height. What is the ratio of the volume of the cylinder to the volume of the cone? A) 2:1 B) 1:2 C) 3:1 D) 1:3

37. The area of a circle is 49π square inches. What is its circumference? A) 7π inches B) 14π inches C) 21π inches D) 49π inches

38. What is the area of a regular hexagon with side length 4 units? A) 24 units² B) 24√3 units² C) 48 units² D) 96 units²

39. The base of a pyramid is a square with side length 12 cm. If the volume of the pyramid is 192 cm³, what is its height? A) 4 cm B) 8 cm C) 12 cm D) 16 cm

40. A sector of a circle has radius 10 cm and arc length 5π cm. What is the central angle in degrees? A) 90° B) 60° C) 45° D) 30°

41. What is the volume of a rectangular prism with dimensions 4 units × 6 units × 5 units? A) 60 units³ B) 120 units³ C) 150 units³ D) 200 units³

42. The surface area of a sphere is 100π square units. What is its volume? A) $250\pi/3$ units³ B) $125\pi/3$ units³ C) $500\pi/3$ units³ D) $1000\pi/3$ units³

43. A regular pentagon has a side length of 6 cm. What is the apothem (the perpendicular distance from the center to any side)? A) 3 cm B) 4.13 cm C) 5.23 cm D) 6.18 cm

44. If the edge length of a cube is increased by 50%, by what percentage does its volume increase? A) 50% B) 125% C) 150% D) 237.5%

45. The total surface area of a cylinder with height 10 cm is 180π cm². What is the radius of the cylinder? A) 3 cm B) 5 cm C) 6 cm D) 9 cm

46. A cone has a height of 12 cm and a slant height of 15 cm. What is the radius of its base? A) 5 cm B) 7 cm C) 9 cm D) 13 cm

47. The areas of two similar triangles are 16 cm² and 36 cm². What is the ratio of their corresponding sides? A) 4:9 B) 2:3 C) 16:36 D) 4:6

48. Which expression represents the volume of a sphere with radius r? A) $4\pi r^2$ B) $4\pi r^3/3$ C) πr^3 D) $2\pi r^3$

49. A parallelogram has sides of lengths 10 cm and 12 cm, with one angle measuring 30°. What is its area? A) 60 cm² B) $60\sqrt{3}$ cm² C) 120 cm² D) $30\sqrt{3}$ cm²

50. The length, width, and height of a rectangular prism are in the ratio 3:2:1. If the volume is 48 cubic units, what is the total surface area? A) 36 units² B) 48 units² C) 64 units² D) 88 units²

ANSWERS

Foundations of Geometry

1. B - Line is an undefined term in geometry, along with point and plane; circles, triangles, and angles are defined using these terms.

2. A - Using the distance formula: $d = \sqrt{(-2-4)^2 + (5-(-3))^2} = \sqrt{36+64} = \sqrt{100} = 10$.

3. A - A ray has one endpoint and extends infinitely in one direction.

4. A - Using the midpoint formula: $\left((3+(-5))/2, (-7+9)/2\right) = (-1,1)$.

5. A - Points are collinear if the distance from A to C equals the sum of distances AB and BC.

6. C - Adjacent angles share a common vertex and a common side.

7. B - When parallel lines are cut by a transversal, alternate interior angles are congruent.

8. B - Angle 5 is the corresponding angle to angle 1, so it equals 65°.

9. A - The point P(3, 4) satisfies the equation $3 \times 3 + 4 \times 4 = 9 + 16 = 25$, not 12, so the distance is not 0.

10. C - Using the distance formula: $d = \sqrt{(5-(-1))^2 + (4-2)^2} = \sqrt{36+4} = \sqrt{40} = 2\sqrt{10}$.

11. A - Complementary angles sum to 90°.

12. A - The midpoint divides a line segment into two congruent parts.

13. C - The parallel postulate states through a point not on a line, exactly one line can be drawn parallel to the given line.

14. A - Vertical angles are congruent.

15. B - Using the section formula: $\left((2 \times 8 + 1 \times 2)/(2+1), (2 \times 9 + 1 \times (-3))/(2+1)\right) = (18/3, 15/3) = (6,5)$.

16. C - A protractor is used to measure angles.

17. C - Trisecting an arbitrary angle is impossible using only compass and straightedge (proven in the 19th century).

18. B - Point (-5, 7) has negative x and positive y, placing it in quadrant II.

19. C - $Slope = (5-(-1))/(7-3) = 6/4 = 3/2$.

20. A - From the origin, 3 units right is (3, 0), and 4 units up gives (3, 4).

21. B - Perpendicular lines have slopes whose product is -1.

22. C - $Distance = \sqrt{(2^2 + 3^2)} = \sqrt{(4 + 9)} = \sqrt{13}$.

23. A - perpendicular line Construction shows the correct method for a perpendicular line Construction using compass arcs of equal radius from both endpoints.

24. A - If M(3, -2) is the midpoint of PQ and P(5, 1) is one endpoint, then $Q(2 \times 3 - 5, 2 \times (-2) - 1) = (1, -5)$.

25. C - Parallel lines can be vertical, in which case both have undefined slopes (not equal slopes).

26. A - In the standard form $(x - h)^2 + (y - k)^2 = r^2$, the center is at (h, k), so x = 3.

27. B - Complementary angles sum to 90°, so the complement is 90° - 25° = 65°.

28. D - Point (4, -5) has positive x and negative y, placing it in quadrant IV.

29. A - The perpendicular bisector passes through the midpoint of the segment and is perpendicular to it.

30. B - The midpoint formula is $((a + c)/2, (b + d)/2)$.

31. A - Points (2, 3), (4, 7), and (6, 11) all lie on the line $y = 2x - 1$.

32. B - This is a statement of the parallel postulate in Euclidean geometry.

33. A - The distance between (a, b) and (a, c) is the absolute difference of the y-coordinates: |b - c|.

34. A - Perpendicular lines intersect at a 90° angle.

35. A - Horizontal lines have a slope of 0.

36. A - The center of a circle is the midpoint of any diameter, so $center = ((2 + 8)/2, (3 + 9)/2) = (5,6)$.

37. B - Finding the midpoint (4, 3) and using the negative reciprocal of the slope of the original segment gives y = -x + 8.

38. C - Constructing a regular pentagon requires techniques beyond basic compass and straightedge constructions.

39. C - Triangle is a defined term (a polygon with three sides); point, line, and plane are undefined.

40. B - If corresponding angles are congruent when lines are cut by a transversal, the lines are parallel.

41. A - The standard form for a circle with center (h, k) and radius r is $(x - h)^2 + (y - k)^2 = r^2$, so $(x - 2)^2 + (y - (-3))^2 = 4^2$.

42. A - Point (3, 0) is 3 units away from the origin since $d = \sqrt{(3^2 + 0^2)} = 3$.

43. A - Congruent angles have the same measure.

44. B - If the product of two slopes is $-1 (2 \times (-1/2) = -1)$, the lines are perpendicular.

45. A - An angle bisector divides an angle into two congruent angles.

46. C - A line segment AB is correctly denoted with a bar above: $\overline{AB}$.

47. A - If midpoint is (3, 4) and one endpoint is (1, 7), then other endpoint is $(2 \times 3 - 1, 2 \times 4 - 7) = (5, 1)$.

48. A - Point (3, 2) lies on the line y = 2, so the distance is 0.

49. B - Complementary angles sum to 90°, so angle B = 90° - 37° = 53°.

50. B - Points on the x-axis always have a y-coordinate of 0.

Logical Reasoning and Proof

1. A - The converse of "If P, then Q" is "If Q, then P," so the converse of "If it rains, then the grass is wet" is "If the grass is wet, then it rains."

2. B - A conditional statement "If p, then q" is false only when the hypothesis p is true and the conclusion q is false.

3. B - Inductive reasoning moves from specific observations to general principles, while deductive reasoning moves from general principles to specific conclusions.

4. B - A conditional and its contrapositive always have the same truth value, while a conditional and its inverse always have opposite truth values.

5. C - The contrapositive of "If P, then Q" is "If not Q, then not P," so the contrapositive is "If a triangle does not have three congruent angles, then it does not have three congruent sides."

6. D - A proof by counterexample demonstrates that a statement is false by finding a specific case where the statement fails to hold true.

7. B - In a two-column proof, the left column contains statements, and the right column provides the reasons or justifications for each statement.

8. B - Using the transitive property with "All squares are rectangles" and "All rectangles have four sides," we can conclude that "All squares have four sides."

9. A - Angle 8 is alternate exterior to angle 1, and since the lines are parallel, these angles are congruent, making angle 8 measure 65° (which is 180° - 115°).

10. C - Vertical angles are always congruent (equal in measure).

11. B - When two lines are cut by a transversal and alternate exterior angles are congruent, the lines are parallel.

12. C - This is a valid deductive argument following the form: All A are B; C is an A; therefore, C is a B.

13. C - A biconditional statement combines a conditional and its converse with "if and only if," meaning both parts must have the same truth value.

14. B - A paragraph proof presents the logical argument in continuous prose form, while a two-column proof organizes the argument into parallel columns of statements and reasons.

15. C - When parallel lines are cut by a transversal, consecutive interior angles are supplementary (sum to 180°).

16. B - The inverse of "If P, then Q" is "If not P, then not Q," so the inverse is "If an animal is not a dog, then it does not have four legs."

17. B - Deductive reasoning, which moves from general principles to specific conclusions, is most commonly used in geometric proofs.

18. A - "All right angles are congruent" is a postulate (or axiom) in Euclidean geometry, which is accepted without proof.

19. B - A flow proof uses boxes with statements connected by arrows to show the logical flow of the argument.

20. C - If angles 1 and 2 are complementary, and angles 2 and 3 are complementary, then angles 1 and 3 must be congruent (equal in measure).

21. C - AAA (Angle-Angle-Angle) is not a valid criterion for proving triangles congruent; it only proves similarity.

22. C - This syllogism follows the valid form: All A are B; All C are A; therefore, All C are B.

23. A - The contrapositive "If I don't pass the test, then I didn't study" is logically equivalent to the original statement "If I study, then I will pass the test."

24. D - In a triangle, the sum of angles is 180°, so $\angle B = 180° - 40° - 60° = 80°$.

25. B - In a proof by contradiction, reaching a contradiction after assuming the original statement is true proves that the original statement must be false.

26. D - Congruent triangles have corresponding parts congruent, but there's no rule that AB must equal BC within the same triangle.

27. C - When parallel lines are cut by a transversal, corresponding angles are congruent.

28. B - Axioms are accepted without proof as a starting point, while theorems are proven using axioms, definitions, and previously proven theorems.

29. C - The contrapositive "If I am not in the United States, then I am not in Florida" is logically equivalent to the original statement.

30. A - In a valid argument, if the premises are true, the conclusion must also be true.

31. B - One way to prove a quadrilateral is a parallelogram is to show that two pairs of opposite sides are congruent.

32. B - The Law of Detachment states that if p→q is true and p is true, then q must be true.

33. C - The negation of "All A are B" is "Some A are not B," so the negation is "Some triangles do not have three sides."

34. A - Using the transitive property of conditionals: If A→B and B→C, then A→C, so "If it is a square, then it has four sides."

35. B - Corresponding angles of congruent triangles are congruent, which is a valid reason in a proof.

36. D - Proof by authority (accepting something as true because an authority says so) is not a valid method of mathematical proof.

37. C - Without knowing the truth value of q, we cannot determine the truth value of "If p, then r".

38. B - Placing a vertex at the origin simplifies calculations by eliminating some coordinates (making them zero).

39. C - A rectangle has four sides, but they are not all congruent, so it serves as a counterexample to the statement.

40. B - The sum of interior angles in an n-sided polygon is (n-2)×180°, so for a pentagon (n=5), the sum is (5-2)×180° = 540°.

41. C - If the angles are in ratio 2:3:4, then the sum must be 180°. If 2x + 3x + 4x = 180°, then 9x = 180°, so x = 20°, making the angles 40°, 60°, and 80°.

42. A - Inductive reasoning draws general conclusions from specific observations, as in observing many rectangles and concluding all have four sides.

43. B - The symbol "≅" represents congruence in geometry.

44. C - Showing all sides have the same length proves a quadrilateral is a rhombus, not necessarily a parallelogram.

45. A - The transitive property states that if a = b and b = c, then a = c.

46. C - "Through any three points, there is exactly one circle" is not a postulate about lines and points in Euclidean geometry; in fact, three collinear points don't determine a circle.

47. C - An exterior angle equals the sum of the remote interior angles, so the other angle is 125° - 45° = 80°.

48. C - The law of contrapositive states that a conditional statement $(p \rightarrow q)$ is logically equivalent to its contrapositive $(\sim q \rightarrow \sim p)$.

49. B - In a flow proof, arrows represent the logical connections between statements, showing how each statement follows from previous ones.

50. B - An indirect proof (proof by contradiction) assumes the conclusion is false and derives a contradiction, thus proving the original conclusion.

Transformations

1. D - Both translations and reflections are rigid transformations that preserve both size and shape.

2. C - The rule $(x, y) \rightarrow (x + 3, y - 5)$ describes a translation 3 units right and 5 units down.

3. A - Reflection over the x-axis changes only the y-coordinates: $(x, y) \rightarrow (x, -y)$.

4. A - Translations always produce congruent figures because they preserve both distance and angle measure.

5. B - A reflection reverses the orientation of a figure (changes it from clockwise to counterclockwise or vice versa).

6. A - Reflection over the y-axis changes the sign of the x-coordinate: $(x, y) \rightarrow (-x, y)$.

7. B - The rule for a 90° counterclockwise rotation around the origin is $(x, y) \rightarrow (-y, x)$.

8. B - Under a dilation with scale factor 2, lengths are doubled (so perimeter doubles) and areas are multiplied by $2^2 = 4$.

9. C - The triangle appears to have been rotated 180° around the origin, mapping quadrant I to quadrant III.

10. D - Dilation is not a rigid transformation because it changes the size of the figure.

11. A - Reflection over the x-axis changes the sign of the y-coordinate: $(x, y) \rightarrow (x, -y)$.

12. C - Under a translation, a line always maps to a parallel line. Rotations and reflections generally change the direction.

13. A - Under a dilation with scale factor 3 from the origin, the center (2, 3) becomes (6, 9) and the radius 4 becomes 12.

14. A - A rotation of 180° around the origin is equivalent to reflecting over both the x and y axes.

15. B - The composition of two reflections over parallel lines results in a translation perpendicular to the lines.

16. C - A non-identity translation has no fixed points; all points move the same distance in the same direction.

17. B - Reflection over the line y = x switches the x and y coordinates: (x, y) → (y, x).

18. D - Dilations (except with scale factor 1) change the area of a figure by a factor of k^2.

19. C - A 180° rotation around the origin changes the signs of both coordinates: (x, y) → (-x, -y).

20. A - A figure with rotational symmetry of order 4 can be rotated by 90° to coincide with itself.

21. C - Under a dilation with center O and scale factor k, the distance from O to any point is multiplied by k.

22. A - The rule for reflection over the y-axis is (x, y) → (-x, y).

23. A - Translating (-2, 3) by vector <5, -7> gives (-2+5, 3+(-7)) = (3, -4).

24. D - Dilations (except with scale factor 1) do not preserve distances between points.

25. C - A figure with one line of reflectional symmetry has rotational symmetry of order 2 (it maps onto itself after a 180° rotation).

26. B - The composition of reflections over the x-axis and y-axis is equivalent to a 180° rotation around the origin.

27. C - The hexagon has doubled in size while maintaining the same center, indicating a dilation with scale factor 2.

28. C - A circle has infinite lines of reflectional symmetry and rotational symmetry of infinite order.

29. B - Under a dilation with scale factor k, the area of a figure is multiplied by k^2.

30. C - A rectangle is invariant (maps onto itself) under reflection over a line through its center parallel to a side.

31. C - A translation never has fixed points (unless it's the identity transformation).

32. B - The rule (x, y) → (-y, x) represents a 90° counterclockwise rotation around the origin.

33. D - A dilation with scale factor 2 would quadruple the area: $16 \times 2^2 = 64$.

34. C - Any rigid transformation (translation, rotation, or reflection) can map a figure onto itself if the figure has the appropriate symmetry.

35. B - After a 270° counterclockwise rotation around the origin: (5, 2) → (2, -5).

36. B - A reflection over a line maps the line to itself because every point on the line is fixed.

37. C - A quadrilateral with rotational symmetry of order 2 must have diagonals that bisect each other.

38. B - Two figures are congruent if and only if one can be mapped to the other using rigid transformations.

39. A - Composing dilations with scale factors 3 and 1/3 gives a scale factor of 3 × (1/3) = 1, which is the identity transformation.

40. C - Reflecting (4, -3) over the line y = x gives (-3, 4) by exchanging and negating coordinates.

41. D - A dilation with different scale factors in different directions (a non-uniform scaling) changes the shape.

42. B - The composition of two reflections over lines that intersect at angle θ is equivalent to a rotation of 2θ around their point of intersection.

43. C - A figure with rotational symmetry of order 6 can be rotated by $360°/6 = 60°$ to map onto itself.

44. A - Under a dilation with scale factor 2, all lengths (including perimeter) are doubled.

45. D - All the transformations mentioned (translation, reflection, rotation) have inverse transformations.

46. C - The line y = x is the perpendicular bisector of the segment joining (3, 5) and (5, 3).

47. B - After reflection over the x-axis, a line with slope m has slope -m.

48. A - A rigid transformation (isometry) preserves collinearity, distances, and angle measures.

49. C - A rotation of 180° around the origin maps (a, b) to (-a, -b).

50. D - A dilation from the origin with scale factor 3 multiplies all coordinates by 3, mapping the given triangle as described.

Congruence and Similarity

1. C - SSA (Side-Side-Angle) is not a valid triangle congruence criterion as it can result in two different triangles.

2. C - An isosceles triangle has at least two congruent sides.

3. B - With a similarity ratio of 2:3, the perimeter ratio is also 2:3, so perimeter of $DEF = 12 \times (3/2) = 18$ units.

4. C - Congruent triangles have the same size and shape but may have different orientations.

5. C - Calculating the side lengths shows they are all different, making it a scalene triangle.

6. A - By the Pythagorean theorem: $7^2 + 24^2 = 49 + 576 = 625 = 25^2$, confirming it's a right triangle.

7. D - If the scale factor is 3, the area ratio is $3^2 = 9$, so the ratio is 9:1.

8. A - AAA (Angle-Angle-Angle) is valid for similarity because if all angles are congruent, the triangles are similar.

9. B - The orthocenter is the point of intersection of the three altitudes of a triangle.

10. B - The triangles share two sides and the included angle (SAS): AB = DE, BC = EF, and angle B = angle E.

11. B - If the ratio of corresponding sides is 2:5, the ratio of areas is $(2/5)^2 = 4/25$.

12. C - Each angle in an equilateral triangle measures 60°.

13. B - A midsegment connects the midpoints of two sides of a triangle.

14. B - In similar triangles, corresponding sides are proportional: AB/DE = BC/EF = AC/DF.

15. C - The circumcenter is the point of intersection of the perpendicular bisectors of the sides of a triangle.

16. C - In a right triangle, the sum of angles is 180°, so 90° + 30° + x = 180°, giving x = 60°.

17. B - In an isosceles triangle with two congruent sides, the angles opposite those sides are also congruent.

18. D - Having one angle of 60° doesn't guarantee an isosceles triangle; it could be in any type of triangle.

19. C - Congruent triangles have the same area, so triangle DEF also has an area of 24 square units.

20. C - The Triangle Inequality Theorem states that the sum of any two sides must be greater than the third side.

21. C - A 30°-60°-90° triangle has three different angles and three different sides, making it scalene.

22. C - In an equilateral triangle, the centroid, orthocenter, circumcenter, and incenter all coincide at the center.

23. A - By the angle bisector theorem, $AB/AC = 4/5$, so $AC = (AB \times 5)/4 = (6 \times 5)/4 = 7.5$.

24. B - Reflection is a rigid transformation that preserves size and shape.

25. A - Using Heron's formula or the formula $A = (1/2)bh$ with b = 3 and h = 4, the area is 6 square units.

26. A - The triangles share two pairs of angles (AA): angle P = angle S and angle Q = angle T, which is sufficient for similarity.

27. D - In a 45°-45°-90° triangle, if the hypotenuse is 10, each leg is $10/\sqrt{2} = 5\sqrt{2} \approx 7.07 units$.

28. D - The three medians of a triangle intersect at the centroid.

29. B - SAS congruence requires two pairs of sides and the included angle to be congruent.

30. A - If the area ratio is 9:16, the ratio of corresponding sides is √(9/16) = 3/4.

31. D - The incenter is the point of intersection of the angle bisectors of a triangle.

32. A - The largest side (AB = 10) is opposite to the largest angle.

33. C - If the ratio of sides is 2:3, the ratio of areas is (2/3)² = 4/9. So if the smaller triangle has area 12, the larger has area 12 × (9/4) = 27.

34. C - AAS (Angle-Angle-Side) is a valid criterion for triangle congruence.

35. A - If one acute angle is twice the other, and their sum is 90°, then $x + 2x = 90°$, giving $x = 30°$ and $2x = 60°$.

36. C - The centroid divides each median in a 2:1 ratio from the vertex to the midpoint.

37. B - Three congruent parts (in the correct combination) are needed to prove triangle congruence.

38. B - In a 30°-60°-90° triangle, if the shortest side is 5, the hypotenuse is $2 \times 5 = 10\ units$.

39. A - By the basic proportionality theorem, a line that divides two sides of a triangle proportionally is parallel to the third side.

40. A - The midsegment of a triangle is parallel to the third side and equal to half its length.

41. A - In an isosceles triangle, the altitude to the base also bisects the base.

42. C - The circumcenter is equidistant from the three vertices of a triangle.

43. A - If the perimeter is 24 and two sides are 8 and 10, then the third side is $24 - 8 - 10 = 6\ units$.

44. B - The shortest distance from a point to a line is along the perpendicular from the point to the line.

45. D - Checking with the Pythagorean theorem: $8^2 + 15^2 = 64 + 225 = 289$, but $17^2 = 289$, so it's the right triangle.

46. B - The sum of angles in a triangle is 180°, so the third angle is 180° - 55° - 65° = 60°.

47. B - The triangle has two sides of equal length (9), making it isosceles.

48. D - The lengths of the medians alone do not provide enough information to determine the perimeter.

49. B - With a similarity ratio of 3:4, DF = AC × (4/3) = 6 × (4/3) = 8.

50. B - Using the formula $A = (1/2)|x^1(y^2 - y^3) + x^2(y^3 - y^1) + x^3(y^1 - y^2)|$ or $A = (1/2)bh = (1/2) \times 3 \times 4 = 6\ square units$.

Polygons and Quadrilaterals

1. C - Using the formula (n-2) × 180° for a hexagon (n=6): (6-2) × 180° = 4 × 180° = 720°.

2. C - The defining property of a parallelogram is that its diagonals bisect each other.

3. D - For a regular octagon (n=8): $((n-2) \times 180°)/n = ((8-2) \times 180°)/8 = (6 \times 180°)/8 = 1080°/8 = 135°$.

4. B - In a parallelogram, opposite angles are congruent and consecutive angles are supplementary, so $2x + 10 = 3x - 20$, which gives x = 30.

5. B - For a regular polygon, each exterior angle = 360°/n, so 24° = 360°/n, which gives n = 15.

6. C - In a rhombus, the diagonals are always perpendicular to each other.

7. D - Perpendicular diagonals are a property of rhombuses and kites, not all parallelograms.

8. A - The sum of angles in a quadrilateral is 360°, so 75° + 95° + 110° + x = 360°, which gives x = 80°.

9. B - Area of a rhombus = $(d_1 \times d_2)/2 = (8 \times 12)/2 = 48\ cm^2$.

10. C - For a regular pentagon (n=5), each exterior angle = 360°/n = 360°/5 = 72°.

11. D - The diagonals of a square form four congruent triangles (not false).

12. C - A trapezoid is defined as a quadrilateral with exactly one pair of parallel sides.

13. C - Perimeter of a regular hexagon = 6 × side length = 6 × 5 = 30 cm.

14. B - The sum of exterior angles of any convex polygon is always 360°.

15. B - The midsegment of a trapezoid equals the average of the bases: (8 + 14)/2 = 11 cm.

16. C - A rhombus does not necessarily have all angles congruent; only opposite angles are congruent.

17. C - A quadrilateral with four congruent sides is a rhombus (may or may not have right angles).

18. B - The interior angle sum formula for an n-sided polygon is (n-2) × 180°.

19. C - The formula for the number of diagonals is n(n-3)/2, so 9(9-3)/2 = 9(6)/2 = 27.

20. B - In an isosceles trapezoid, the diagonals are congruent.

21. B - Area of a trapezoid = (1/2) × height × (sum of bases) = (1/2) × 4 × (5 + 11) = 2 × 16 = 32 cm².

22. B - The formula for the number of diagonals in an n-sided polygon is n(n-3)/2.

23. C - Using the formula for interior angles: ((n-2) × 180°)/n = 140°, solving gives n = 9.

24. C - The diagonals of a kite are perpendicular to each other.

25. A - In a regular hexagon, the side length equals the radius of the circumscribed circle.

26. B - The area of a regular hexagon with side length s is $(3\sqrt{3}/2) \times s^2 = (3\sqrt{3}/2) \times 4^2 = 24\sqrt{3}$ square units.

27. D - A square is both a rhombus (all sides congruent) and a rectangle (all angles are right angles).

28. C - Sum of interior angles in a pentagon is $(5 - 2) \times 180° = 540°, so\ 100° + 110° + 120° + 130° + x = 540°$, giving x = 80°.

29. C - Perpendicular diagonals are a property of rhombuses, not all rectangles.

30. C - For a regular 20-sided polygon: $((n - 2) \times 180°)/n = ((20 - 2) \times 180°)/20 = (18 \times 180°)/20 = 3240°/20 = 162°$.

31. B - In a parallelogram, opposite angles are congruent, so the other angles are 70° and 110°.

32. C - A quadrilateral with both pairs of opposite sides parallel is, by definition, a parallelogram.

33. C - A regular hexagon has 6 lines of symmetry.

34. C - For a rhombus with diagonals 10 cm and 24 cm, the side length is $\sqrt{(5^2 + 12^2)} = 13cm$, so $perimeter = 4 \times 13 = 52\ cm$.

35. C - Only in a square are the diagonals perpendicular bisectors of each other.

36. A - The area of a regular polygon is (1/2) × perimeter × apothem = (1/2) × n × s × a.

37. A - A quadrilateral with one pair of opposite sides that are both parallel and congruent must be a parallelogram.

38. B - The apothem of a regular hexagon with side length s is $\left(\sqrt{3}/2\right) \times s = \left(\sqrt{3}/2\right) \times 8 = 4\sqrt{3}$ cm.

39. D - The diagonals of a trapezoid have no special properties in general.

40. C - A quadrilateral with four congruent sides and four right angles is a square.

41. B - A kite has two pairs of adjacent sides that are congruent.

42. C - Sum of interior angles for a 15-sided polygon: (15-2) × 180° = 13 × 180° = 2340°.

43. A - The midsegment of a trapezoid is parallel to the bases.

44. A - Area of a rhombus = $(side\ length)^2 \times sin(angle) = 5^2 \times sin(60°) = 25 \times (\sqrt{3}/2) = 25\sqrt{3}/2$ cm².

45. B - A quadrilateral with diagonals that bisect each other and are perpendicular must be a rhombus.

46. A - The minimum number of sides for any polygon is 3 (a triangle).

47. B - For a regular octagon (n=8), each exterior angle = 360°/n = 360°/8 = 45°.

48. D - A regular heptagon (7 sides) cannot be constructed using only compass and straightedge.

49. A - Connecting all diagonals in a convex n-sided polygon forms $n - 2$ triangles.

50. B - If the diagonals of a parallelogram are perpendicular, it must be a rhombus.

Right Triangles and Trigonometry

1. B - Use the Pythagorean theorem: $8^2 + b^2 = 17^2$; $b^2 = 289 - 64 = 225$; $b = 15$ units.

2. C - In a 30°-60°-90° triangle, the hypotenuse equals twice the shortest side; $2 \times 5 = 10$ units.

3. A - Using the Pythagorean identity $sin^2\theta + cos^2\theta = 1$: $cos^2\theta = 1 - (3/5)^2 = 1 - 9/25 = 16/25$; $cos\theta = 4/5$.

4. B - Using the Pythagorean theorem: $h^2 = 20^2 - 12^2 = 400 - 144 = 256$; $h = 16$ feet.

5. A - The area of a right triangle is $(1/2) \times base \times height = (1/2) \times 5 \times 12 = 30$ square units.

6. A - If $\tan\theta = 4/3$, then $\sin\theta = opposite/hypotenuse = 4/5$ (using the Pythagorean theorem to find the hypotenuse).

7. B - In a 45°-45°-90° triangle, the hypotenuse equals $leg \times \sqrt{2}$; $6 \times \sqrt{2} = 6\sqrt{2}$ units.

8. A - Using the Pythagorean theorem: $x^2 + 6^2 = 10^2$; $x^2 + 36 = 100$; $x^2 = 64$; $x = 8$.

9. B - If $\cos\theta = 5/13$, then the adjacent side is 5 and the hypotenuse is 13; using the Pythagorean theorem, the opposite side is 12; $\tan\theta = opposite/adjacent = 12/5$.

10. C - By the Pythagorean theorem: $7^2 + 24^2 = 49 + 576 = 625 = 25^2$, so it's a right triangle.

11. D - To be a Pythagorean triple, $a^2 + b^2 = c^2$; $7^2 + 8^2 = 49 + 64 = 113 \neq 11^2 = 121$.

12. A - Using tangent: $\tan\theta = opposite/adjacent$; $\tan 32° = height/50$; $height = 50 \times \tan 32° \approx 31\ meters$.

13. A - If $\sin\theta = 3/5$ for one angle, then $\cos\theta = 4/5$ for that angle; for the other acute angle (90° - θ), $cos = sin = 3/5$.

14. A - Using the Pythagorean identity: $\cos^2\theta = 1 - \sin^2\theta = 1 - (12/13)^2 = 1 - 144/169 = 25/169$; $\cos\theta = 5/13$.

15. B - In a 30°-60°-90° triangle, the side opposite to 30° is (hypotenuse/2); $16/2 = 8$ units.

16. B - If $\sin A = \sin B$, then either $A = B$ or $A = 180° - B$; in either case, sides a and b are equal, so $b = 7$.

17. $B - \tan 45° = 1$ is an exact value to memorize.

18. B - In a right triangle, the acute angles are complementary (sum to 90°); $90° - 25° = 65°$.

19. A - Using sine: $\sin\theta = opposite/hypotenuse = 8/17$ (where $hypotenuse = \sqrt{(8^2 + 15^2)} = 17$).

20. A - Using the Law of Cosines: $\cos C = (a^2 + b^2 - c^2)/(2ab) = (36 + 64 - 100)/(2 \times 6 \times 8) = 0$; $C = 90°$.

21. B - Using the Pythagorean theorem: $wire^2 = 50^2 + 30^2 = 2500 + 900 = 3400$; $wire = \sqrt{3400} \approx 58.3\ feet$.

22. A - To form a right triangle, $a^2 + b^2 = c^2$, $3^2 + 4^2 = 9 + 16 = 25$; $x = 5$.

23. A - If $\cos\theta = 7/25$, then the adjacent side is 7 and the hypotenuse is 25; the opposite side is 24; $\tan\theta - 24/7$.

24. B - Using the Pythagorean theorem: $5^2 + x^2 = (x + 2)^2$; $25 + x^2 = x^2 + 4x + 4$; $25 = 4x + 4$; $x = 5.25$.

25. B - Using sine: $sin\theta = opposite/hypotenuse = 12/24 = 1/2; \theta = 30°$.

26. B - $sin30° = 1/2$ is an exact value to memorize.

27. A - $tan\theta = 3/4; \theta = tan^{-1}(3/4) \approx 37°$.

28. A - $cos60° = 1/2$ is an exact value to memorize.

29. C - Using the Pythagorean theorem: $a^2 + b^2 = c^2; a^2 + 8^2 = 10^2; a^2 + 64 = 100; a^2 = 36; a = 6$ units.

30. D - Using the Law of Cosines to find the angle opposite side 13 gives approximately 120°.

31. A $- sin\theta = opposite/hypotenuse = 15/25 = 3/5$.

32. C - $cos45° = \sqrt{2}/2$ is an exact value to memorize.

33. B - When the Law of Sines gives two possible triangles, it's called the ambiguous case (SSA).

34. B - In a right triangle, the acute angles sum to 90°; 90° - 40° = 50°.

35. A - In a 45°-45°-90° triangle, the ratio of a leg to the hypotenuse is $1: \sqrt{2}$.

36. C - Using the triangle inequality theorem and checking with the Law of Cosines, the triangle is acute.

37. B - The geometric mean of two positive numbers a and b is $\sqrt{(ab)}$.

38. C - Using the sine addition formula: $sin(A + B) = sinA \cdot cosB + cosA \cdot sinB = (4/5)(12/13) + (3/5)(5/13) = 63/65$.

39. C - The sum of angles in any triangle must equal 180°; 30° + 60° + 100° = 190° > 180°.

40. B - $tan\theta$ = opposite/adjacent is the definition of tangent.

41. A - Using Heron's formula with semi-perimeter $s = (13 + 14 + 15)/2 = 21$: $Area = \sqrt{(21(21 - 13)(21 - 14)(21 - 15))} = \sqrt{(21 \cdot 8 \cdot 7 \cdot 6)} = 84$.

42. D - The Pythagorean theorem: $a^2 + b^2 = c^2$ is always true for right triangles.

43. C - Using the Pythagorean theorem: $x^2 + (2x)^2 = c^2; x^2 + 4x^2 = 5x^2; c = x\sqrt{5}$.

44. C - The altitude to the hypotenuse creates two triangles that are similar to each other and to the original triangle.

45. C - Since sine values must be between -1 and 1, sin θ = 1.5 has no solution.

46. B - Using the Law of Sines: $b/sinB = a/sinA; b = (a \cdot sinB)/sinA = (10 \cdot sin45°)/sin30° = 10 \cdot (\sqrt{2}/2)/(1/2) = 10\sqrt{2} \approx 14.1$.

47. B - The area formula for a triangle using two sides and the included angle is $A = (1/2) \cdot ab \cdot sinC$.

48. A - Using sine: $sin60° = height/15$; $height = 15 \cdot sin60° = 15 \cdot (\sqrt{3}/2) \approx 13\ feet$.

49. A - The correct form of the Law of Cosines is $a^2 = b^2 + c^2 - 2bc \cdot cosA$.

50. C - Using the Law of Cosines to find the angle opposite side 9 gives approximately 75°.

Circles

---

1. B - The standard form is $(x-h)^2 + (y-k)^2 = r^2$, where (h,k) is the center and r is the radius.

2. A - The circumference is $2\pi r = 2\pi(6) = 12\pi$ cm.

3. A - Using the arc length formula: $s = r\theta$ where θ is in radians. $8\pi = r(72°)(\pi/180°)$, solving gives r = 20 cm.

4. B - Area of sector = $(\theta/360°)\pi r^2 = (60°/360°)\pi(12)^2 = (1/6)144\pi = 144\pi/3$ cm².

5. B - Completing the square: $(x^2 + 4x + 4) + (y^2 - 6y + 9) = -9 + 4 + 9 = 4$, so center is (-2, 3).

6. A - Using the Pythagorean theorem: $10^2 = 8^2 + x^2$, so x = 6 cm.

7. A - Area = $\pi r^2 = \pi(7)^2 = 49\pi$ cm².

8. B - The measure of an arc equals the measure of its central angle, so arc AB = 130°.

9. A - A tangent line is perpendicular to the radius at the point of tangency.

10. B - The distance from (0, 0) to (0, -7) is 7, so the radius is 7 and the equation is $x^2 + y^2 = 49$.

11. A - Congruent chords in a circle are equidistant from the center.

12. A - Using the distance formula: $r = \sqrt{[(7-4)^2 + (3-(-1))^2]} = \sqrt{[9 + 16]} = \sqrt{25} = 5$.

13. A - In the standard form $(x-h)^2 + (y-k)^2 = r^2$, the radius r is 6.

14. B - The measure of an inscribed angle is half the measure of its intercepted arc.

15. B - The standard form of a circle equation is $(x-h)^2 + (y-k)^2 = r^2$.

16. A - For arc length $s = r\theta$, if $\theta = 1$ radian and s = 12 cm, then r = 12 cm.

17. B - The perpendicular bisector of any chord passes through the center of the circle.

18. A - Area of segment = area of sector - area of triangle = $(90°/360°)\pi(10)^2$ $(1/2)(10)^2sin90° = 25\pi - 50$ cm².

19. A - For externally tangent circles, the distance between centers equals the sum of the radii.

20. B - A secant line intersects a circle at exactly two points.

21. C - If the diameter is 16, the radius is 8, so the equation is $x^2 + y^2 = 64$.

22. B - The line containing the points of intersection is perpendicular to the line connecting the centers.

23. A - If chords AB and CD intersect at point P, then AP·PB = CP·PD (power of a point).

24. B - If PA is tangent and PBC is secant, then PA² = PB·PC.

25. A - Using the Pythagorean theorem: 13² = (12/2)² + d², so d = 5 cm.

26. B - Tangent segments from the same external point to a circle are equal, so PB = PA = 8 cm.

27. A - If C = 18π, then r = 9 cm, so A = πr² = 81π cm².

28. A - Completing the square: (x²-4x+4) + (y²+6y+9) = 12+4+9 = 25, so radius is 5.

29. A - An inscribed angle that intercepts a semicircle is a right angle (90°).

30. D - Area of ring = π(7² - 3²) = π(49 - 9) = 40π cm².

31. D - If the distance between centers equals the sum of radii (5+12=17), the circles are tangent externally.

32. A - Completing the square: $(x^2 - 6x + 9) + (y^2 + 8y + 16) = -15 + 9 + 16 = 10$, so center is (3, -4).

33. D - A line can intersect a circle at exactly one point without being tangent (e.g., at a point of tangency with another curve).

34. B - For a tangent from external point: $t^2 = d^2 - r^2 = 15^2 - 9^2 = 225 - 81 = 144$, so t = 12 cm.

35. C - Area of sector $= (\theta/2\pi)\pi r^2 = (3\pi/4)/(2\pi) \cdot \pi \cdot 6^2 = (3/8) \cdot 36\pi = 27\pi/2$ cm².

36. B - Circles with the same center are concentric circles.

37. C - Point (1, 7) is at distance 5 from (-2, 3): $\sqrt{[(1 - (-2))^2 + (7 - 3)^2]} = \sqrt{[9 + 16]} = 5$.

38. D - The general form is $x^2 + y^2 + Dx + Ey + F = 0$.

39. D - Points on a circle centered at the origin must satisfy x² + y² = r² for some constant r. For (4, 4), x² + y² = 32.

40. A - The distance from (2, -3) to (0, 0) is $\sqrt{(2^2 + (-3)^2)} = \sqrt{13}$, so the equation is $(x - 2)^2 + (y + 3)^2 = 13$.

41. B - The measure of an inscribed angle is half the measure of its intercepted arc: 140°/2 = 70°.

42. B - Chord length $= 2r \cdot sin(\theta/2) = 2 \cdot 8 \cdot sin(30°) = 16 \cdot (1/2) = 8$ cm.

43. A - The angle between a tangent and secant from an external point equals the inscribed angle intercepting the same arc.

44. A - The center is the midpoint of the diameter: $((2 + (-4))/2, (5 + 1)/2) = (-1, 3)$. Radius $= \sqrt{(2 - (-1))^2 + (5 - 3)^2} = \sqrt{9 + 4} = \sqrt{13}$, so the equation is $(x + 1)^2 + (y - 3)^2 = 13$.

45. B - The circumference of a circle is $2\pi r$.

46. D - Arc length $= (\theta/360°) \cdot 2\pi r = (45°/360°) \cdot 2\pi \cdot 10 = (1/8) \cdot 20\pi = 10\pi/4$ cm.

47. A - The perpendicular from the center of a circle to a chord bisects the chord.

48. A - Completing the square: $(x^2 - 4x + 4) + (y^2 + 6y + 9) = -4 + 4 + 9 = 9$, so center is (2, -3) and distance to origin is $\sqrt{(2^2 + (-3)^2)} = 5$.

49. D - Two intersecting circles can have a maximum of 4 common tangent lines (2 external and 2 internal).

50. C - Using the Pythagorean theorem: $r^2 = d^2 + (chord/2)^2 = 12^2 + 5^2 = 144 + 25 = 169$, so r = 13 cm.

Coordinate Geometry

---

1. D - Using the distance formula: $d = \sqrt{(3 - (-2))^2 + (-4 - 5)^2} = \sqrt{25 + 81} = \sqrt{106}$.

2. A - If midpoint is (5, -2) and one endpoint is (8, 1), then other endpoint $= (2 \times 5 - 8, 2 \times (-2) - 1) = (2, -5)$.

3. A - Using point-slope form: $y - (-4) = 3(x - 2)$, so $y + 4 = 3x - 6$, thus $y = 3x - 10$.

4. B - Product of slopes $= -2 \times (1/2) = -1$, so the lines are perpendicular.

5. C - Slope $= (9 - 3)/(5 - 5) = 6/0$, which is undefined. Vertical lines have undefined slope.

6. A - Solving for y: $-2y = -3x + 12$, so $y = (3/2)x - 6$.

7. A - Area $= (1/2) \times$ base $\times$ height $= (1/2) \times 4 \times 4 = 8$ square units.

8. B - Standard form: $(x - h)^2 + (y - k)^2 = r^2$, so $(x - 3)^2 + (y - (-2))^2 = 5^2$.

9. D - Point (4, -3) has positive x and negative y, placing it in quadrant IV.

10. A - Point-slope form: $y - y_1 = m(x - x_1)$, so $y - 5 = -3(x - (-2)) = -3(x + 2)$.

11. C - Two lines are perpendicular if and only if the product of their slopes is -1.

12. B - Distance $= |3(2) + 4(3) - 25|/\sqrt{(3^2 + 4^2)} = |6 + 12 - 25|/5 = |-7|/5 = 7/5 = 1.4$, which rounds to closest option 2.

13. A - If midpoint is (3, -4) and one endpoint is (7, 2), then other endpoint $= (2 \times 3 - 7, 2 \times (-1) - 2) = (-1, -10)$.

14. A - Perpendicular slope $= -1/(2) = -1/2$. Using point-slope: $y - (-1) = -1/2(x - 4)$, so $y + 1 = -1/2x + 2$, thus $y = -1/2x + 1$.

161

15. B - Midpoint of a diameter is the center: $((1 + 7)/2, (2 + 8)/2) = (4, 5)$.

16. A - Distance $= \sqrt{(3^2 + 4^2)} = \sqrt{(9 + 16)} = \sqrt{25} = 5$.

17. A - When x=0: -5y=10, so y=-2. Y-intercept is (0, -2).

18. B - The figure has perpendicular sides with lengths 6 and 4, making it a rectangle.

19. D - If line has equation 4x+3y=12, its slope is -4/3. Perpendicular slope = -1/(-4/3) = 3/4 = -3/4.

20. B - If A to C is 12 units, and B is 1/3 of the way from A to C, then AB = (1/3) × 12 = 4 units.

21. A - Slope-intercept form y = mx + b indicates slope is 3 and y-intercept is (0, -7).

22. C - Distance from (-1, 2) to (3, -6) is $\sqrt{\left(3 - (-1)\right)^2 + (-6 - 2)^2} = \sqrt{16 + 64} = \sqrt{80}$, so equation is $(x+1)^2 + (y - 2)^2 = 80$.

23. A - If (4, -3) is midpoint of A(p, q) and B(r, s), then (p+r)/2 = 4, so p+r = 8.

24. C - Using section formula: $((2 \times 8 + 1 \times 2)/(2 + 1), (2 \times 5 + 1 \times (-3))/(2 + 1)) = (18/3, 7/3) = (6, 2⅓)$, closest to (6, 2).

25. A - Both lines have the same slope -2/3 (from 2x+3y=6 → y=(-2/3)x+2), but different y-intercepts, so they're parallel.

26. A - Completing the square: $(x^2 - 6x + 9) + (y^2 + 8y + 16) + 9 - 9 - 16 = 0, so (x - 3)^2 + (y + 4)^2 = 16$, with center (3, -4).

27. A - Original line's slope is 3/2. Line 6x-4y=10 can be written as $y = (3/2)x - 2.5$, showing it has the same slope.

28. A - Using the formula $Area = \left|(1/2)\left(x^1(y^2 - y^3) + x^2(y^3 - y^1) + x^3(y^1 - y^2)\right)\right| = \left|(1/2)\left(1(7 - (-1)) + 5((-1) - 3) + 3(3 - 7)\right)\right| = 10$.

29. B - Midpoint is (4, 7), slope of original segment is 1, so perpendicular slope is -1. Equation is $y - 7 = -1(x - 4)$, or $y = -x + 11$.

30. A - Standard form $(x - h)^2 + (y - k)^2 = r^2$ indicates radius is $\sqrt{49} = 7$.

31. D - Distance between parallel lines $ax + by + c_1 = 0$ and $ax + by + c_2 = 0$ is $|c^2 - c^1|/\sqrt{(a^2 + b^2)} = |(-12) - 6|/\sqrt{(2^2 + (-3)^2)} = 18/\sqrt{13}$.

32. B - Slope between first two points is $(5 - (-1))/(k - 3) = 6/(k - 3)$. Slope between first and third points is $(9 - (-1))/(7 - 3) = 10/4 = 2.5$. Setting equal: $6/(k - 3) = 2.5$, so $k - 3 = 6/2.5 = 2.4$, thus k = 5.4, closest to 5.

162

33. A - Standard form for ellipse: $(x - h)^2/a^2 + (y - k)^2/b^2 = 1$, with center (h,k)=(2,-3), semi-major axis a=5, and semi-minor axis b=3.

34. A - Using the general form $ax + by + c = 0$ with the two intercepts gives 2x-y=6.

35. A - Using distance formula: $\sqrt{((7 - 3)^2 + (10 - k)^2)} = 10, so (10 - k)^2 = 100 - 16 = 84$, thus $k = 10 \pm \sqrt{84} \approx 10 \pm 9.17 \approx 0.83$ or 19.17, closest to 2 or 14.

36. A - In a parallelogram, diagonals bisect each other. If the intersection point is (3, 2) and three vertices are given, the fourth vertex is $(2 \times 3 - 1, 2 \times 2 - 5) = (5, -1)$.

37. C - The perpendicular bisector of the segment joining (3, 5) and (7, 1) has slope -1.

38. A - A parabola with vertex at (2, -3) and axis parallel to y-axis has the form y = a(x-2)² - 3.

39. A - Standard form $(x - h)^2/a^2 - (y - k)^2/b^2 = 1$ indicates center at (h,k) = (4,-2).

40. D - The slopes of the three sides of a triangle can have any relationship to each other.

41. C - Distance from point to line $ax + by + c = 0$ is $|ax^1 + by^1 + c|/\sqrt{(a^2 + b^2)} = |2(-1) + 1(3) + 5|/\sqrt{(2^2 + 1^2)} = |1|/\sqrt{5} \approx 0.45$, closest to 1.

42. B - For parabola $y = -2(x - 3)^2 + 4$, the vertex is (3, 4) and the focus is $1/(4a) = 1/(4 \times (-2)) = -1/8$ units below the vertex.

43. D - The equation $|x - 3| + |y + 2| = 5$ represents a rhombus with vertices at (3, 3), (8, -2), (3, -7), and (-2, -2).

44. B - Rewriting as x²/6 + y²/4 = 1 gives semi-major axis √6 and semi-minor axis 2.

45. A - The tangent line to a circle at point (x_1, y_1) has equation $x_1 x + y_1 y = r^2$, so $3x + 4y = 25$.

46. C - Using the Shoelace formula (or coordinate method) gives area = 16 square units.

47. A - Reflection across y=x swaps the coordinates, so (-2, 5) reflects to (5, -2).

48. C - Distance between centers = $\sqrt{((4 - 1)^2 + (6 - 2)^2)} = \sqrt{(9 + 16)} = 5$. Since $3 + 5 = 8 > 5$ and $|5 - 3| = 2 < 5$, the circles intersect.

49. A - Eccentricity of an ellipse = $\sqrt{(1 - b^2/a^2)} = \sqrt{(1 - 9/25)} = \sqrt{(16/25)} = 4/5$.

50. B - Completing the square: $(x^2 + 6x + 9) + (y^2 - 8y + 16)$ | 25 - 9 - 16 - 0, so $(x + 3)^2 + (y - 4)^2 = 0$, which is a point (degenerate circle).

Geometric Measurement and Dimension

1. B - Perimeter of rectangle $= 2(length + width) = 2(12 + 7) = 2(19) = 38$ cm.

2. B - Using $C = 2\pi r$, we get $24\pi = 2\pi r$, so $r = 12\ cm$.

3. C - Perimeter of regular hexagon $= 6 \times side\ length = 6 \times 5 = 30$ units.

4. B - Area of triangle $= (1/2) \times base \times height$, so $36 = (1/2) \times 12 \times h$, giving h = 6 units.

5. A - Area of circle $= \pi r^2$, where r = d/2 = 5 m, so area $= \pi(5)^2 = 25\pi$ m².

6. B - Area of trapezoid $= (1/2) \times height \times (sum\ of\ parallel\ sides) = (1/2) \times 6 \times (8 + 14) = 3 \times 22 = 66$ cm².

7. B - Area of parallelogram $=$ base $\times$ height, so $120 = b \times 8$, giving b = 15 units.

8. A - Area of sector $= (\theta/360°) \times \pi r^2 = (60°/360°) \times \pi(12)^2 = (1/6) \times 144\pi = 24\pi$ cm².

9. D - Volume of cube $= s^3 = 5^3 = 125$ in³.

10. B - Volume of rectangular prism $= 1 \times$ w $\times$ h, so 360 = 10 $\times$ 6 $\times$ h, giving h = 6 units.

11. A - Volume of triangular prism $=$ area of base $\times$ height $= (1/2) \times 3 \times 4 \times 10 = 60$ cm³.

12. B - Volume of cylinder $= \pi r^2 h = \pi \times 5^2 \times 8 = \pi \times 25 \times 8 = 200\pi$ m³.

13. C - Surface area of cube $= 6s^2 = 6 \times 4^2 = 6 \times 16 = 96$ units².

14. B - Volume of cone $= (1/3) \times \pi r^2 h = (1/3) \times \pi \times 6^2 \times 8 = (1/3) \times \pi \times 36 \times 8 = 96\pi$ cm³.

15. A - Surface area of sphere $= 4\pi r^2$, so $256\pi = 4\pi r^2$, giving r² = 64 and r = 8 (incorrect); actual answer is r = 8.

16. B - Volume of pyramid $= (1/3) \times$ base area $\times$ height $= (1/3) \times 8 \times 6 \times 10 = (1/3) \times 480 = 160$ cm³.

17. B - Lateral surface area of cylinder $= 2\pi rh = 2\pi \times 3 \times 7 = 42\pi$ m².

18. C - Volume of sphere $= (4/3)\pi r^3 = (4/3)\pi \times 5^3 = (4/3)\pi \times 125 = 500\pi/3$ cm³.

19. A - Volume of cone $= (1/3)\pi r^2 h$, so $24\pi = (1/3)\pi \times 3^2 \times$ h, giving h = 8 units.

20. D - For a square with perimeter 36, each side = 36/4 = 9 units, so area = 9² = 81 units².

21. B - Area of rhombus $= (1/2) \times$ product of diagonals $= (1/2) \times 10 \times 16 = 80$ cm².

22. D - When all dimensions are doubled, volume increases by a factor of $2^3 = 8$.

23. B - For a regular pentagon with perimeter 45, each side = 45/5 = 9 units.

24. B - Surface area $= 2(lw + lh + wh) = 2(5 \times 3 + 5 \times 4 + 3 \times 4) = 2(15 + 20 + 12) = 2(47) = 94$ cm².

25. D - When radius is doubled, volume increases by a factor of $2^3 = 8$.

26. C - Surface area of cylinder = $2\pi r^2 + 2\pi rh = 2\pi \times 4^2 + 2\pi \times 4 \times h = 32\pi + 8\pi h$; Setting this equal to 96π: $32\pi + 8\pi h = 96\pi$, solving gives h = 8 inches.

27. C - Perimeter of regular octagon = 8 × side length = 8 × 4 = 32 units.

28. A - Using Heron's formula with $s = (5 + 12 + 13)/2 = 15$, $area = \sqrt{(15 \times 10 \times 3 \times 2)} = 30$ cm².

29. A - Volume of sphere = $(4/3)\pi r^3 = (4/3)\pi \times 3^3 = (4/3)\pi \times 27 = 36\pi$ units³.

30. B - Volume = base area × height; Using Heron's formula for the triangle area with s = 15, area = 30 cm², so volume = 30 × 9 = 270 cm³.

31. C - For a square with area 121, side length = $\sqrt{121} = 11$, so perimeter = 4 × 11 = 44 units.

32. D - Perimeter = 2(8) + 4 + (4 + π×4) = 16 + 4 + 4 + 4π = 24 + 4π cm.

33. B - Lateral surface area = (1/2) × perimeter × slant height = (1/2) × (4×6) × 5 = (1/2) × 24 × 5 = 60 units².

34. C - For a cube with surface area 96, $6s^2 = 96$, so $s^2 = 16$ and s = 4; Volume = $s^3 = 4^3 = 64$ units³.

35. A - Volume of cone = $(1/3)\pi r^2 h$, so $40\pi = (1/3)\pi \times r^2 \times 15$, giving $r^2 = 8$ and $r = 2\sqrt{2}$ cm (incorrect); actual answer is r ≈ 2.83 cm.

36. C - The ratio is 3:1 because the volume of a cylinder is 3 times the volume of a cone with the same base and height.

37. B - For a circle with area 49π, $r^2 = 49$, so r = 7; Circumference = $2\pi r = 2\pi \times 7 = 14\pi$ inches.

38. B - Area of regular hexagon = $\left(3\sqrt{3}/2\right) \times s^2 = \left(3\sqrt{3}/2\right) \times 4^2 = \left(3\sqrt{3}/2\right) \times 16 = 24\sqrt{3}$ units².

39. A - Volume of pyramid = $(1/3) \times base\ area \times height = (1/3) \times 12^2 \times h = 48h = 192$, giving h = 4 cm.

40. A - Arc length = $(\theta/360°) \times 2\pi r$, so $5\pi = (\theta/360°) \times 2\pi \times 10$, giving $\theta = 90°$.

41. B - Volume of rectangular prism = l × w × h = 4 × 6 × 5 = 120 units³.

42. C - For a sphere with surface area 100π, $4\pi r^2 = 100\pi$, so $r^2 = 25$ and r = 5; Volume = $(4/3)\pi r^3 = (4/3)\pi \times 5^3 = (4/3)\pi \times 125 = 500\pi/3$ units³.

43. B - For a regular pentagon with side length 6, the apothem is approximately 4 13 cm (using trigonometry).

44. D - New volume = $(1.5)^3 \times$ original volume = 3.375 × original volume, an increase of 237.5%.

45. C - Surface area = $2\pi r^2 + 2\pi rh = 2\pi r(r + h) = 180\pi$; With h = 10, we have $2\pi r(r + 10) = 180\pi$, so r(r + a10) = 90, solving gives r = 6 cm.

46. C - Using the Pythagorean theorem in a right triangle, $r^2 + h^2 = s^2$, so $r^2 = s^2 - h^2 = 15^2 - 12^2 = 225 - 144 = 81$, giving $r = 9$ cm.

47. B - For similar triangles, the ratio of areas equals the square of the ratio of corresponding sides; $\sqrt{16/36} = \sqrt{4/9} = 2/3$.

48. B - The volume of a sphere with radius r is $(4/3)\pi r^3$.

49. A - Area of parallelogram = base × height = $10 \times 12 \times \sin(30°) = 120 \times 0.5 = 60$ cm².

50. D - If the ratio is 3:2:1 and volume is 48, then dimensions are 6×4×2 = 48; Surface area = 2(6×4 + 6×2 + 4×2) = 2(24 + 12 + 8) = 2(44) = 88 units².

Made in United States
Orlando, FL
02 July 2025

62554023R10094